Urban Sustainability

The Urban Sustainability Book Series is a valuable resource for sustainability and urban-related education and research. It offers an inter-disciplinary platform covering all four areas of practice, policy, education, research, and their nexus. The publications in this series are related to critical areas of sustainability, urban studies, planning, and urban geography.

This book series aims to put together cutting-edge research findings linked to the overarching field of urban sustainability. The scope and nature of the topic are broad and interdisciplinary and bring together various associated disciplines from sustainable development, environmental sciences, urbanism, etc. With many advanced research findings in the field, there is a need to put together various discussions and contributions on specific sustainability fields, covering a good range of topics on sustainable development, sustainable urbanism, and urban sustainability. Despite the broad range of issues, we note the importance of practical and policy-oriented directions, extending the literature and directions and pathways towards achieving urban sustainability.

The series will appeal to urbanists, geographers, planners, engineers, architects, governmental authorities, policymakers, researchers of all levels, and to all of those interested in a wide-ranging overview of urban sustainability and its associated fields. The series includes monographs and edited volumes, covering a range of topics under the urban sustainability topic, which can also be used for teaching materials.

Jiangbo Wang · Kai Liu · Tao Liu ·
Toshiyuki Yamamoto

Customized Bus: User Travel Behavior and Demand Evolution

Springer

Jiangbo Wang
Dalian University of Technology
Dalian, Liaoning, China

Kai Liu
Dalian University of Technology
Dalian, Liaoning, China

Tao Liu
Southwest Jiaotong University
Chengdu, Sichuan, China

Toshiyuki Yamamoto
Nagoya University
Nagoya, Aichi, Japan

ISSN 2731-6483 ISSN 2731-6491 (electronic)
Urban Sustainability
ISBN 978-981-96-3487-3 ISBN 978-981-96-3488-0 (eBook)
https://doi.org/10.1007/978-981-96-3488-0

This Springer imprint is published by the registered company Springer Nature Singapore Pte Ltd.
The registered company address is: 152 Beach Road, #21-01/04 Gateway East, Singapore 189721, Singapore

Contents

Part I
Customized Bus

Chapter 1
An Overview of Customized Bus System

Abstract The Customized Bus (CB) system is a demand-responsive transit model aimed at addressing critical urban mobility challenges such as traffic congestion, environmental concerns, and insufficient public transport coverage in low-demand areas. Introduced in China in 2013, the CB system utilizes advanced digital platforms to design flexible routes and schedules tailored to groups with similar travel demands. This chapter examines the operational characteristics, user behavior, and market orientation of CB services, highlighting their potential as a complementary solution to traditional public transport (PT) and a sustainable alternative to private vehicles. Despite offering personalized and efficient transit services, CB systems encounter challenges in cost management, demand forecasting, and sustaining user loyalty. Understanding spatial–temporal demand patterns, passenger loyalty determinants, and the influence of the built environment on travel behaviors is crucial for optimizing the effectiveness and sustainability of the Customized Bus system.

Keywords Customized Bus · Background · Market orientation · Passenger loyalty · Demand evolution

1.1 Background

Demand responsive transportation (DRT) is a form of shared transportation service for clustered travel demand where vehicles can alter their routes based on particular travel demand rather than following a fixed route and timetable (Wikipedia Contributors 2020). It has been widely implemented around the world overpast decades and is seriously considered as a means of providing attractive mobility at a lower overall cost than conventional PT services, albeit at a higher per passenger trip cost (Mulley and Nelson 2009; Nelson and Phonphitakchai 2012; Wang et al. 2014). Supported by the rapid development of information and communication technologies, various types of demand-responsive mobility services such as dial-a-ride and flexible transportation services have been entering the life of the public and contributing to improving the satisfaction of mobility, for instance, for elderly and disabled people and in areas

J. Wang et al., *Customized Bus: User Travel Behavior and Demand Evolution*, Urban Sustainability, https://doi.org/10.1007/978-981-96-3488-0_1

of low demand where regular bus service is not financially viable. Nevertheless, the DRT services have often been criticized owing to their relatively high cost of provision, their lack of flexibility in route planning, their inability to manage high demand, and uncertainty of traveling reliability (such as ride time) (Mageean and Nelson 2003; Nelson and Phonphitakchai 2012). In practice, simply implementing an advanced telematics-based DRT system is not sufficient to address these limitations (Mageean and Nelson 2003). As a type of on-demand system, understanding the traveling behavior of passengers, ascertaining the mechanism of demand generation, and identifying and forecasting the potential demand are critical to the service operation. The emerging techniques such as big data mining and multi-scale data fusion provide a promising solution to these challenges (Ma et al. 2013).

The customized bus (CB) service has rapidly gained popularity in major cities across China in the recent decade, promising to offer more attractive and diverse PT services and thereby reducing urban traffic congestion, especially during peak hours. By the end of 2019, car ownership in China first increased to 260 million (Abigail 2020). The resulting increasing travel demand in urban areas has led to severe traffic congestion and traffic-related fatalities and injuries (Pucher et al. 2007). The continually increasing number of vehicles on the roads also has a significant impact on greenhouse gas emissions and climate change (Chapman 2007). In the face of these dilemmas, local authorities have spared no effort to develop and expand the PT network (Zhang et al. 2018). Developing an advanced, appealing, and diversified PT system is prioritized as a potential solution against these challenges, in conjunction with policies to restrict the use and purchase of private cars, such as the license plate lottery for vehicle purchasing and vehicles plate number traffic restriction (Zhong et al. 2012). The CB service, which is famous for convenience, comfort, and personalized attribute, came into being in this context.

Similar to the DRT systems, the CB route design is also flexible and demand-responsive (as summarized in Table 1.1). With an advanced online travel platform, the CB system aims to provide high-quality and personalized transportation services for a specific group of travelers that have similar spatial–temporal travel requests. Unlike the DRT systems operated in US and European countries, the CB systems developed in China mainly provide bus-sharing transportation services but not focus on mobility of single demand, such as an elderly or disabled person. The CB systems in China have mainly been operated by local transit agencies, which are usually state-owned enterprises and supported and subsided by local governments, aiming to maximize total social benefits (Liu and Ceder 2015). In contrast to the regulation of fixed-route PT services, the CB operators have been authorized with great flexibility in terms of CB route design and operation. They can design and adjust CB routes fully depending on demand, as long as the stations are designed in places where a temporal stop is allowed. CBs are allowed to use dedicated bus lanes and existing bus stops. The DRT services in US and European countries are generally provided by private companies that are selected, authorized, and funded (partially or fully) by local transit authorities (Wikipedia Contributors 2020).

Fixed-route bus systems provide regular and universal mobility services to residents and are generally effective in densely populated areas. However, in low demand

Table 1.1 Similarities and differences between the general DRT and CB systems

	General DRT system	CB system
Route design	Determined by demand; flexible or semi-flexible (with a limited number of fixed stations); updated daily	Determined by demand; semi-flexible or fixed for a period of time (such as a month); updated daily for the daily-ticket routes and adjusted monthly for the monthly-ticket route
Demand characteristics	Real-time demand; a single passenger can also be served; mainly complement PT service to provide mobility for elderly, disabled people and in areas without PT service	Real-time (not so real-time as the general DRT system is) and regular demand; a group of passengers who have similar travel requests (close origin/destination and departure/arrival time); bus-sharing system
Vehicle types used	Medium-/small-sized bus, van, car, taxi	Medium-/small-sized bus, van
Technology	Mainly by telephone calls to make a reservation	Smartphone APP, easy-to-access web, or WeChat Mini Programs passengers can see the location of the bus on an electronic map
Operating entity	Private enterprises with or without subsidies from local transit authorities	Local transit agencies (stated-owned) and cooperate with private enterprises

areas, such as sprawled suburban regions where demand can fluctuate significantly, the flexible route design has been proven to have a lower system cost than the fixed one, even though with additional advertising and promotion costs included (Nourbakhsh and Ouyang 2012). Since the demand for these services is often unpredictable and largely depends on the established service itself, determining the optimal demand level or the conditions that would warrant a shift from fixed-route to flexible-route transit service is not straightforward (Quadrifoglio and Li 2009). Using an operation simulation framework (Nourbakhsh and Ouyang 2012), investigated the relation ship between demand level and system costs and found that the flexible-route transit system is advantageous when demand is low-to-moderate (i.e., 4–40 passengers per hour per km^2). In the study areas of this dissertation, 97% of the one-kilometer cells have fewer than 40 passengers per hour served by the CB service (more than 78% of the cells with 4–40 passengers per hour served by the CB services).

The rationale of the CB service is illustrated in Fig. 1.1. Particularly, the CB aggregates users with similar origins, destinations, and departure/arrival times, which can be clustered by demarcating the demand distribution into OD bundles. Each OD bundle is composed of an origin zone, a destination zone, and a passing-by corridor. The ladder-shaped origin/destination demarcations are beneficial for shortening the detour distance by picking up and dropping off passengers. The boarding

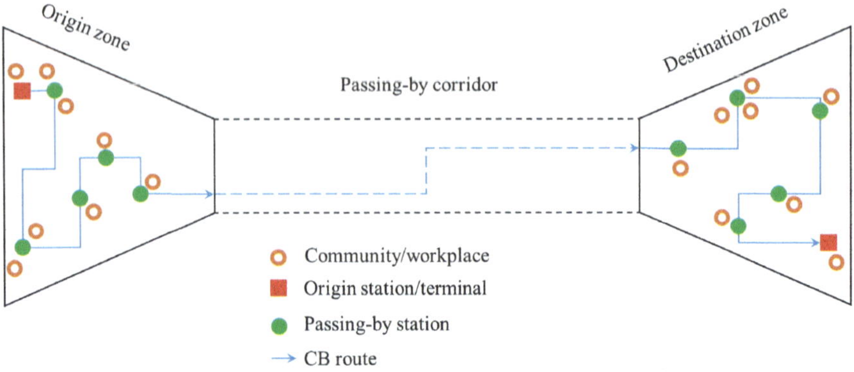

Fig. 1.1 Typical example of a CB route

and alighting stations are designed to locate close to the users' residences and workplaces. Furthermore, the number of stations within the passing corridor is kept to a minimum to enhance route flexibility, while drivers are guided by the vehicle navigation system to ensure they follow the shortest and most time-efficient paths. Travelers can subscribe to CB services by using easy-to-access websites, smartphone applications, and WeChat mini programs. The CB service has rapidly become popular in China due to its simplicity, convenience, comfort, and customized features. It is now operated in more than 70 cities in China. As of June 2024, Beijing boasts 1244 customized bus routes, serving a daily ridership of up to 66,000 passengers, making it the largest CB service in China.

The concept of the CB service is based on car-sharing systems, which originated in Zurich in 1948 and operated until 1998 (Harms and Truffer 1998; Shaheen and Cohen 2007). The earliest studies concerning the implementation of the car-sharing concept into the PT system were conducted in the 1970s (Kirby and Bhatt 1974; Morin 1974). Henceforth, several subscription transit services sprung up across the US (Bautz 1975; Kirby and Bhatt 1975) and some European countries (Davison et al. 2014) to compensate for the PT service in areas with poor or nonexistent PT service. It was the rudiment of demand-responsive and user-oriented transit service, known as paratransit/flexible transit service or dial-a-ride service.

The development of these DRT services underlies the appearance of the CB service. In August 2013, a new personalized transit service first appeared in Qingdao, China (Zhan and Dong 2014). Since then, the CB services have become popular across China (Liu and Ceder 2015) and other countries such as Singapore (Abdullah 2017). Different from the conventional PT services, CB services aim at providing higher-quality transit services (albeit at a higher cost) for a relatively small number of passengers with similar travel demand (Liu and Ceder 2015). Therefore, the design and operation process of a CB can be considered to be a combination of a conventional PT system with a demand-driven bus-sharing system and subscription concept (Liu et al. 2016).

1.2 Challenges in the Initial Adoption and Sustainability of CB

Several months after its prevalence, there was a drastic decline in the number of CB service users in some cities of China such as Guangzhou, Hangzhou, Nanjing, Xiamen, and Harbin (Chu 2015; Ma 2015). The resulting cancelation of launching routes and user subscriptions brought a huge waste of resources. If we understand the following reasons, the decline in subscriptions and cancelation of launching routes in these cases seem inevitable in the progress of CB service:

- As an emerging transport mode, planners had not even understood the role of CB in the transportation system but just blindly followed its prevalence. Meanwhile, travelers do not have an image about the CB trips and therefore are not aware of whether it is appropriate personally;
- For developing CB systems, the effects and mechanisms of frailties and factors on subscribing behavior are still not clear yet;
- There is a lack of CB demand forecasting techniques, which hinders the understanding of the spatiotemporal evolution of CB demand.

1.3 Market Orientation of CB Services

Particularly, the orientation of CB services (or namely the role of CB services) in the market of the transportation system determines its primary service object, system construction, and government function. As a demand-driven, user-oriented transit service, the travel requests of passengers have been put as a top priority (Cao and Wang 2016; Liu and Ceder 2015; Zhang et al. 2017). The design and operation of CB service must be prioritized to meet the specific needs of passengers. Blindly following trends in CB system development without a clear understanding of their role in the broader transportation system can lead to inefficient route designs and poor service quality.

From the perspective of travelers, it is difficult to understand the actual performance of the CB service before a trial, for there are always some discrepancies between the business advertisements and actual performance. Travelers might not even recognize the differences between using a CB service and their previous mode of transport without a practical trial.

The blind spread of CB systems has led to the operation of a large number of low-efficiency CB routes, and those novelty-seeking travelers have blindly followed the trend of subscribing to the services with poor service quality, jointly inducing the lose-lose situation in the development of CB services. Therefore, it is imperative to explore the role of CB services based on an investigation into the practical performance, advantages and disadvantages compared to other modes of the transportation system, not only for the development of the service system but also to provide prospective users important references in mode choice.

1.4 CB Traveling Behavior

As the long-term passenger subscription is vital to the survival and viability of the CB system, a better understanding of passenger behavioral loyalty toward the CB service is beneficial for improving the systematic operation efficiency and decreasing the probability of wasting resources due to service termination. As the travel requests of passengers are assigned as a top priority, identifying loyal passengers and targeting with intensive strategies to retain them is a win–win solution for both operators and those loyal users. The drastic decline in subscriptions immediately after routes are launched indicates the high-level frailties in loyalty of passengers toward the services, such as the novelty-seeking behavior of some travelers. Therefore, the urgent and outstanding issue is to ascertain which frailty and factor determine the long-term subscription behavior and how they function.

Feedback from the viewpoint of user subscription behavior is a good indicator of user loyalty toward the supplied service and will provide practical references for policy-making and marketing direction (Jomnonkwao et al. 2015; Wen et al. 2005). Factors leading to the frail loyalty can be targeted with intensive strategies to maximize user retention.

Additionally, the built environment has been discovered to significantly impact travel behavior, and therefore planners have bent themselves to shaping residents' travel behavior, by changing the built environment, into a healthier and environment-friendly way (Bohte et al. 2009; Cao et al. 2009; Ewing and Cervero 2010).

To understand the CB subscribing behavior of users, the study on the subscriber loyalty pays more attention to the factors directly contributing to the service quality, while an investigation into the built environment is more of revealing the causality behind. For example, if a traveler does not own a private car, a considerable distance from PT stations may result in a higher reliance on CB services. Ascertaining the effects of the built environment on subscribing behavior is beneficial for understanding the potential principles in their decision-making.

1.5 Demand Evolution Patterns

Travel demand has been found to exhibit extensive spatial dependence theoretically and empirically, particularly in ridesharing systems (Becker et al. 2017; Friedrich and Noekel 2017; Ke et al. 2017; Ni et al. 2018). As with other ridesharing systems, a CB route potential occurs, if the temporal and spatial travel requests of a certain number of demanders match (Friedrich et al. 2018). More specifically, CB routing generally addresses time window constraints in picking up and dropping off passengers (Liu and Ceder 2015; Tong et al. 2017). With the expected boarding and arrival times of passengers matched, a CB can actively absorb passengers from areas it passes through, and therefore the spatial dependence of CB demand occurs. The spatial dependence effects are more significant in areas close to the origins and destinations of CB routes because the CBs are permitted to detour more in these areas to maximize ridership. However, travel demand may continue to exhibit the same trend as historical demand (Ke et al. 2017; Liu et al. 2019; Yao et al. 2018), a phenomenon that is generally regarded as time dependence. These characteristics pose challenges in CB demand modeling because of the need to simultaneously address various spatial effects and temporal adjustments.

As a demand-driven transit system, the first step of service design is to collect potential demand with travel surveys (Liu and Ceder 2015). However, the travel requests gathered from stated preference surveys primarily reflect the willingness of individuals to try the CB service based on the service quality advertised by providers, rather than indicating their actual choices. Therefore, to ensure the efficiency of the CB system and the win–win solution for service operators and loyal users, the travel survey alone is far from enough. The lessons drawn from those failed CB operation cases drive us to attach more attention to the actual demand evolution over space and time simultaneously.

The loyalty analysis and investigation on the effects of the built environment provide some empirical references for demand forecasting from individual perspectives. The investigation on the spatiotemporal evolution of CB demand will further provide empirical policy suggestions for service operation and demand identification from a more macro perspective. In this part, this study aims to model the CB demand and shed light on spatial dependence effect, temporal dependence effect, spatial spillover effect, and effect of unit-specific heterogeneity in the demand using a spatial dynamic panel model that can accommodate these effects simultaneously.

1.6 Structure of the Content

The content of this book is structured as follows. As for the first part, Chapter 2 presents an investigation on the actual performance of a successful CB system to shed light on the role and orientation of CB service in the transportation system and provide prospective users with important references in mode choice. In Chapter 3, we constructed a methodological framework to quantify potential benefits of CB services for commuting trips, focusing on aspects such as travel cost, travel time, and fuel consumption in two western cities.

Part II mainly focuses on the travel behavior of CB users. Faced with various challenges in the operation of CB services in several cities across China, including low ridership, difficulty in balancing operational flexibility with stable traveler demand. Chapter 4 aims to explore the mechanism behind CB service choices considering the diversity in travelers' needs and their underlying psychological factors. To better retain continuous users to the maximum extent, in Chapter 5, we further investigated the mechanism of continuous choice behavior of CB passengers based on a survey targeting long-term CB users. To develop targeted and differentiated marketing strategies accordingly, in Chapter 6, we distinguished the active users and inactive users and investigated their differences in subscribing behavior through a comparative analysis, based on a long-term continuous observation on a successful CB system. In Chapter 7, survival analysis is conducted to investigate passenger loyalty and quantify the effects of determinants that directly contribute to the service quality.

Part III aims to present the effects of the built environment on the survival of CB routes and the evolution of demand. Chapter 8 demonstrates the significant impacts of the built environment on the long-term demand stability and potential markets for CB systems through quantitative assessments. In Chapter 9, we further investigated the nonlinear and threshold relationships of built environment variables with CB demand, controlling for individual characteristics and service features. In Chapter 10, we empirically investigate spatial dependence and spillover effects in CB demand using spatial dynamic panel models, revealing that changes in demand rather than the level of demand itself drive spatial dependencies.

References

Abdullah, Z. 2017. Customised bus services set to go on trial in 2018 [WWW Document]. *The Straits Times*. URL https://www.straitstimes.com/singapore/transport/customised-bus-services-set-to-go-on-trial-in-2018. Accessed 26 Dec 20.

Abigail, 2020. China's private car parc exceeds 200 mln for first time [WWW Document]. URL http://autonews.gasgoo.com/m/Detail/70016758.html. Accessed 19 Feb 20.

Bautz, J.A. 1975. Subscription service in the United States. *Transportation (Amst)* 4: 387–402. https://doi.org/10.1007/BF00174738.

Becker, H., F. Ciari, and K.W. Axhausen. 2017. Modeling free-floating car-sharing use in Switzerland: A spatial regression and conditional logit approach. *Transportation Research Part C: Emerging Technologies* 81: 286–299. https://doi.org/10.1016/j.trc.2017.06.008.

Bohte, W., K. Maat, and B. van Wee. 2009. Measuring attitudes in research on residential self-selection and travel behaviour: A review of theories and empirical research. *Transport Reviews* 29: 325–357. https://doi.org/10.1080/01441640902808441.

Cao, X. (Jason), P.L. Mokhtarian, and S.L. Handy. 2009. Examining the impacts of residential self-selection on travel behaviour: A focus on empirical findings. *Transport Reviews*. https://doi.org/10.1080/01441640802539195

Cao, Y., and J. Wang. 2016. The key contributing factors of customized shuttle bus in rush hour: A case study in Harbin city. *Procedia Engineering* 137: 478–486. https://doi.org/10.1016/j.proeng.2016.01.283.

Chapman, L. 2007. Transport and climate change: A review. *Journal of Transport Geography* 15: 354–367. https://doi.org/10.1016/j.jtrangeo.2006.11.008.

Chu, F. 2015. Customized bus: Cherish what you have at the moment. *Transport Business China* 8: 52–55.

Davison, L., M. Enoch, T. Ryley, M. Quddus, and C. Wang. 2014. A survey of demand responsive transport in great Britain. *Transport Policy (Oxf)* 31: 47–54. https://doi.org/10.1016/j.tranpol.2013.11.004.

Ewing, R., and R. Cervero. 2010. Travel and the built environment. *Journal of the American Planning Association* 76: 265–294. https://doi.org/10.1080/01944361003766766.

Friedrich, M., M. Hartl, and C. Magg. 2018. A modeling approach for matching ridesharing trips within macroscopic travel demand models. *Transportation (Amst)* 45: 1639–1653. https://doi.org/10.1007/s11116-018-9957-5.

Friedrich, M., and K. Noekel. 2017. Modeling intermodal networks with public transport and vehicle sharing systems. *EURO Journal on Transportation and Logistics* 6: 271–288. https://doi.org/10.1007/s13676-015-0091-7.

Harms, S., and B. Truffer. 1998. *The emergence of a nation-wide car-sharing co-operative in Switzerland*. Holland.

Jomnonkwao, S., V. Ratanavaraha, B. Khampirat, S. Meeyai, and D. Watthanaklang. 2015. Factors influencing customer loyalty to educational tour buses and measurement invariance across urban and rural zones. *Transportmetrica A: Transport Science* 11: 659–685. https://doi.org/10.1080/23249935.2015.1060274.

Ke, J., H. Zheng, H. Yang, and X. Chen (Michael). 2017. Short-term forecasting of passenger demand under on-demand ride services: A spatio-temporal deep learning approach. *Transportation Research Part C: Emerging Technologies* 85: 591–608. https://doi.org/10.1016/j.trc.2017.10.016

Kirby, R.F., and K.U. Bhatt. 1974. *Guidelines on the operation of subscription bus services*. Washington, DC.

Liu, T., and A. Ceder. 2015. Analysis of a new public-transport-service concept: Customized bus in China. *Transport Policy (Oxf)* 39: 63–76. https://doi.org/10.1016/j.tranpol.2015.02.004.

Liu, T., A.A. Cedera, R. Bologna, and B. Cabantous. 2016. Commuting by customized bus: A comparative analysis with private car and conventional public transport in two cities. *Journal of Public Transportation* 19: 55–74. https://doi.org/10.5038/2375-0901.19.2.4.

Liu, T.L.K., P. Krishnakumari, and O. Cats. 2019. Exploring demand patterns of a ride-sourcing service using spatial and temporal clustering. In 6th International Conference on Models and Technologies for Intelligent Transportation Systems. https://doi.org/10.1109/MTITS.2019.8883312

Ma, X. 2015. Why passengers do not accompany with customized bus? *Decision-Making* 10: 82–83.

Ma, X., Y.J. Wu, Y. Wang, F. Chen, and J. Liu. 2013. Mining smart card data for transit riders' travel patterns. *Transportation Research Part C: Emerging Technologies* 36: 1–12. https://doi.org/10.1016/j.trc.2013.07.010.

Mageean, J., and J.D. Nelson. 2003. The evaluation of demand responsive transport services in Europe. *Journal of Transport Geography* 11: 255–270. https://doi.org/10.1016/S0966-6923(03)00026-7.

Morin, D.A. 1974. Commuter bus clubs are showing the "professionals" how to meet transit demand. *Traffic Engineering, Institute of Traffic Engineering* 44: 24–27.

Mulley, C., and J.D. Nelson. 2009. Flexible transport services: A new market opportunity for public transport. *Research in Transportation Economics* 25: 39–45. https://doi.org/10.1016/j.retrec.2009.08.008.

Nelson, J.D., and T. Phonphitakchai. 2012. An evaluation of the user characteristics of an open access DRT service. *Research in Transportation Economics* 34: 54–65. https://doi.org/10.1016/j.retrec.2011.12.008.

Ni, L., X. Wang (Cara), and X. Chen (Michael). 2018. A spatial econometric model for travel flow analysis and real-world applications with massive mobile phone data. *Transportation Research Part C: Emerging Technologies* 86: 510–526. https://doi.org/10.1016/j.trc.2017.12.002

Nourbakhsh, S.M., and Y. Ouyang. 2012. A structured flexible transit system for low demand areas. *Transportation Research Part B: Methodological* 46: 204–216. https://doi.org/10.1016/j.trb.2011.07.014.

Pucher, J., Z.R. Peng, N. Mittal, Y. Zhu, and N. Korattyswaroopam. 2007. Urban transport trends and policies in China and India: Impacts of rapid economic growth. *Transport Reviews* 27: 379–410. https://doi.org/10.1080/01441640601089988.

Quadrifoglio, L., and X. Li. 2009. A methodology to derive the critical demand density for designing and operating feeder transit services. *Transportation Research Part B: Methodological* 43: 922–935. https://doi.org/10.1016/j.trb.2009.04.003.

Shaheen, S., and A. Cohen. 2007. Growth in worldwide carsharing: An international comparison. *Transportation Research Record: Journal of the Transportation Research Board* 1992: 81–89. https://doi.org/10.3141/1992-10.
Tong, L. (Carol), L. Zhou, J. Liu, and X. Zhou. 2017. Customized bus service design for jointly optimizing passenger-to-vehicle assignment and vehicle routing. *Transportation Research Part C: Emerging Technologies* 85: 451–475. https://doi.org/10.1016/j.trc.2017.09.022
Wang, C., M. Quddus, M. Enoch, T. Ryley, and L. Davison. 2014. Multilevel modelling of Demand Responsive Transport (DRT) trips in Greater Manchester based on area-wide socio-economic data. *Transportation (Amst)* 41: 589–610. https://doi.org/10.1007/s11116-013-9506-1.
Wen, C.H., L.W. Lan, and H.L. Cheng. 2005. Structural equation modeling to determine passenger loyalty toward intercity bus services. *Journal of Transportation Research Board* 1927: 249–255. https://doi.org/10.3141/1927-28.
Wikipedia Contributors. 2020. Demand responsive transport [WWW Document]. Wikipedia. URL https://en.wikipedia.org/w/index.php?title=Demand_responsive_transport&oldid=963553410. Accessed 14 July 2020.
Yao, H., F. Wu, J. Ke, X. Tang, Y. Jia, S. Lu, P. Gong, Z. Li, J. Ye, and D. Chuxing. 2018. Deep multi-view spatial-temporal network for taxi demand prediction. In: 32nd AAAI Conference on Artificial Intelligence, AAAI 2018.
Zhan, H.L., and J.S. Dong. 2014. Qingdao launched the first customized bus in China. *Transportation of Business in China* 9: 20–23.
Zhang, J., D.Z.W. Wang, and M. Meng. 2017. Analyzing customized bus service on a multimodal travel corridor: An analytical modeling approach. *Journal of Transportation Engineering, Part A: Systems* 143: 04017057. https://doi.org/10.1061/jtepbs.0000087.
Zhang, L., R. Long, H. Chen, and T. Yang. 2018. Analysis of an optimal public transport structure under a carbon emission constraint: A case study in Shanghai, China. *Environmental Science and Pollution Research* 25: 3348–3359. https://doi.org/10.1007/s11356-017-0660-4.
Zhong, R., J. Sun, and Q. Lu. 2012. China's public transportation: Problems, policies, and prospective of sustainability. *ITE Journal* 82: 36–40.

Chapter 2
Role and Orientation of Customized Bus Services

Abstract After attaining great prevalence from the end of 2013, customized bus (CB) transit services have experienced a huge decline in China. The feasibility of this new bus sharing system is thus being questioned. Therefore, it is imperative to investigate the actual role of CB services in the overall transportation system based on successful cases, as the role of the CB service determines its primary service object, system construction, marketing orientation, and even government function. To examine the role of CB services, this study investigates the practical performance, advantages, and spatial and temporal coverage of a successful CB system based on practical subscription data for more than two years. The results illustrate that the CB service is an eclectic choice that can balance service quality and cost between traveling by traditional public transportation (PT) and private cars/taxis. Even though the travel cost increased to a limited extent, the CB service significantly improved the travel experience in terms of the travel time, travel speed, number of stations, and arrival time compared to PT services. The multinomial logit model and regression models demonstrate a significantly positive relationship between the relative advantage and amount of demand for the CB services. Furthermore, the CB service primarily serves trips generated during the peak traffic hours of the city and supplements traditional PT service in areas with poor coverage levels.

Keywords Customized Bus · Travel demand analysis · Transit subscription · Performance evaluation · Service quality

2.1 Introduction

The CB service has gained popularity in many large and medium-sized cities in China since its launch in Qingdao in August 2013. It is popular with citizens because of its simplicity, convenience, comfort, and customized features. Despite the prevalence of CB services, drastic declines in CB subscribers appear in some cities, such as Guangzhou, Hangzhou, Nanjing, Xiamen, and Harbin. Subsequently, numerous

J. Wang et al., *Customized Bus: User Travel Behavior and Demand Evolution*, Urban Sustainability, https://doi.org/10.1007/978-981-96-3488-0_2

launched routes have been canceled and many users have terminated subscribing from the service and returned to their previous transport modes.

Blindly following trends in the development of CB systems, without a clear understanding of their role within the broader transportation framework, has resulted in significant resource waste in cities that have seen a drastic decline in passenger numbers. In particular, the role of the CB service will determine its primary service object, system construction, marketing orientation, and even government policy development. On the other hand, from the aspect of passengers, it is difficult to understand the performance of the CB service before a trial, as there are always some discrepancies between the business advertisements and actual performance.

To the best of our knowledge, there is still no consensus on the role of CB services in the transportation system, even though the CB system has been developed for more than 5 years. Therefore, it is imperative to investigate its practical performances, advantages, and disadvantages compared to other modes in the transportation system, not only for the development of the service system but also to provide prospective users with an image of CB travel.

Thus, this chapter seeks to fill this gap by investigating the actual performance of a successful CB system and sheds light on the competitiveness of CB services in the transportation system.

2.2 Literature Review

On account of the lack of studies on the performance evaluation of CB services, it is worthwhile to perform an extensive review of the key studies pertaining to performance evaluation and factor selection for similar transit systems. Kirby and Bhatt (1975) analyzed ten subscription bus services and identified the following characteristics as key to their success: (1) increased focus on long-distance trips, (2) a dedicated organizational body, (3) convenient sources of buses and drivers, (4) high service reliability, (5) high priority of regulatory requirements, (6) continual adjustment, and (7) satisfactory personalized features. Davison et al. (2014) explored demand-responsive transit systems in Great Britain to determine the factors affecting the competitiveness of the systems and their underlying reasons. They highlighted the importance of providing access and geographical coverage, which were also emphasized by Sun et al. (2018), and the effect of the number of seats and vehicles was quantified. Based on attributes in terms of economic viability, service provisions, and technical performance (Mageean and Nelson 2003), evaluated a set of demand-responsive transportation technologies and operations at urban and rural sites across Europe. Zou et al. (2014) established an assessment index system and infrastructure construction, social benefit, and government function were included. Nelson and Phonphitakchai (2012) evaluated the performance of LinkUp, one of the most successful demand-responsive transport systems in the UK, from the viewpoints of awareness, satisfaction, and relative advantage. They pointed out that the demand-responsive transit system has a main function to complement the existing PT services when regular PT services are not operating and where regular PT services are poor. Their results also confirmed that the demand-responsive transport system is very useful to link isolated communities and population groups to the essential services (such as healthcare). Most of these metrics concluded in the literature are from a macro point of view and therefore provide an overview of performance evaluation of the CB systems.

In a more specific way, a series of factors were quantified to evaluate the performance of the PT system. Based on an analysis of the operation of the PT system in Beijing, China, (Li et al. 2013) indicated that average running time and service reliability are the key factors influencing the performance. Similarly, Redman et al. (2013) also highlighted the service reliability as an important factor for attracting car users, and it significantly affected individual perceptions, motivations, and contexts. de Oña et al. (2018) classified the influences of service quality into two underlying dimensions: transport service and comfort and convenience factors. Dell'Olio et al. (2011) found that the wait times, cleanliness, and comfort are valuable attributes for users of PT systems. Liu et al. (2016) compared the performance of CB with private car and PT in Auckland and Paris, using a proposed simulation framework. Differential arrival time, travel time, travel cost, and fuel consumption were calculated to investigate the difference between the travel modes, their viability, and competitiveness based on 100 randomly generated demand for each city. The attributes relating to route and service characteristics, reliability, comfort and cleanliness, fare, information, safety and security, personnel and customer services, and environmental protection were investigated by Eboli and Mazzulla (2011) to measure the performance of PT systems. They also pointed out that a more objective measurement of the metrics can be a useful solution for obtaining a more comprehensive service quality measurement.

2.3 Metrics and Data Collection

To investigate the actual performance of the Dalian CB services (which have been operating successfully for more than five years and providing about 5000 person-time transit services per day), a series of metrics including route and service characteristics, reliability, and travel cost were calculated based on route generation/planning module of online maps and the subscription records.

2.3.1 *Route and Service Characteristics*

The route and service characteristics can be evaluated based on the travel speed and number of stops on the route (Eboli and Mazzulla 2011). Based on the travel demand data obtained in this study, the average travel speed (*ATS*) and the number of stops

(*NOS*) were employed as performance metrics of the route and service characteristics for each OD pair.

The *ATS* is an important measure of performance, and is determined by the following:

$$ATS = {}^{TD}/_{TT} \tag{2.1}$$

where *TD* is the travel distance, and *TT* is the travel time. In this study, the *TD* and *TT* were obtained from the program on the Baidu Map. *TD* and *TT* were calculated for each OD pair for all travel modes. The time and distance required for the user to walk to his/her car, the time required for the user to wait for a taxi or public transit at the origin station, the time required for the user to park his/her car, and the distance traveled for parking were not considered in the calculation of the *TT* and *TD*.

NOS is the total number of stops on the CB/PT routes from the origin to the destination for each passenger. Particularly, the *NOS* for a CB trip was fetched based on the actual route information, while the *NOS* for a PT trip was obtained based on Baidu Map.

2.3.2 Service Reliability

Service reliability reflects the ability of transit service to depart and arrive on time, which is a crucial aspect for users of public transit services (Eboli and Mazzulla 2011). The difference in arrival time is an index that can be used to describe the level of punctuality and reliability of PT services. Liu et al. (2016) used the absolute value of the difference in the arrival time to compare the punctuality of different travel modes. However, whether they are arriving at their destination earlier or later makes a large difference for travelers, especially for commuters. Thus, in this chapter, the difference in arrival time (*DAT*) is determined by the following:

$$DAT = EAT - AAT \tag{2.2}$$

where *EAT* indicates the expected arrival time, and *AAT* denotes the actual arrival time. However, the *EAT*s are not available due to the privacy protection policy of Dalian CB system. According to the principle of CB route design, passengers with similar expected departure and arrival times are aggregated and served by a CB route. Therefore, the designed arrival time of the CB route was approximately regarded as the *EAT* for its passengers. The *AAT* was determined based on the route planning module of the Baidu Map with the designed route departure time input. Particularly, a positive *DAT* indicates that the passenger arrives at his/her destination earlier than the expected arrival time, while a negative *DAT* indicates that the passenger arrives at his/her destination later than the expected arrival time.

As the *DAT* was commonly employed to reflect the reliability of transit service, it was calculated for CB and PT only in this study. Even though we can also calculate

DAT for other modes like private cars and taxis with the same departure time set in the route planning program, it does not have enough representation of the reliability of these modes due to the riders can freely adjust their departure time depending on their actual arrival time and clock-on time. However, for PT and CB, the *DAT* indicates the reliability of their routing and services, given the same departure time.

2.3.3 Travel Cost

The travel cost of a trip, including both monetary cost and time cost, is a major factor that significantly influences a passenger's choice of travel mode (Chowdhury et al. 2015). Thus, travel expense (*TE*) and travel time (*TT*) were employed as travel cost metrics in this study. For PT and taxis, the *TE*s can be fetched from the Baidu Map. As for private cars, the *TE* was calculated as the summation of fuel costs and parking fees, as given by:

$$TE = \frac{C_f \times TD}{100} \times P_f + (T_p \times P_p) \tag{2.3}$$

where C_f is the average fuel consumption per 100 km for passenger vehicles; P_f is the fuel price; T_p is the total parking time, for which half of the total working hours (typically 8 h) was taken as the parking time for each trip; and P_p is the parking fees per hour. It was assumed in this study that all private cars used gasoline. The price of 95-octane gasoline was 7.2 RMB per liter in Dalian on December 14, 2017, according to (Jintounet 2017). The average fuel consumption of passenger vehicles in China reached 6.9 L per 100 km in 2015 according to a report published by the Ministry of Industry and Information Technology of the People's Republic of China (2015). The parking fee in urban areas of Dalian was 5 RMB per hour according to the Price Bureau of Dalian City (2014). Hence, the *TE* for private cars can be determined using Eq. (2.2–2.3).

The travel expenses for CB trips (TE_{CB}) were determined based on the distance tariff scheme formulated (Liu and Ceder 2015) as follows:

$$TE_{CB} = \alpha_1 F_1 + \alpha_2 F_2 \tag{2.4}$$

$$F_1 = max\left\{0, F\prime\left\lceil\frac{TD_{CB} - L_0}{L_1}\right\rceil\right\} + F_0 \tag{2.5}$$

where F_1 is the variable fare; F_2 is the constant fare; α_1 and α_2 are the discount factors for the variable fare and constant fare, respectively; F' is the fare factor for different travel distances; L_0 is the threshold travel distance; L_1 is the travel distance factor; $\lceil x \rceil$ is a function from which the smallest integer greater than or equal to x can be acquired; and F_0 is the base fare for CB services.

Table 2.1 Parameter value used to determine the monthly and daily travel expenses for CB services in Dalian

Parameter	α_1	α_2	F_2	F'	L_0	L_1	F_0
Monthly	0.8	1	0	2 RMB	10 km	5 km	6 RMB
Daily	1	1	0	2 RMB	10 km	5 km	6 RMB

The values of the parameters used to determine the monthly and daily travel subscription tariffs for CB services in Dalian are presented in Table 2.1 based on a previous study conducted by (Liu and Ceder 2015).

2.3.4 Study Area

The study area for this investigation is Dalian, which is a coastal city located in Northeast China. Dalian is a major city in Liaoning Province, China, and plays a leading role in the economic development of the province. The population of Dalian is more than 6.69 million according to the results of the 6th National Population Census conducted in 2010 (National Bureau of Statistics of the People's Republic of China 2011). At the end of 2016, it was reported that the total car ownership in Dalian had reached 1.37 million vehicles (Dalian Municipal Bureau of Statistics 2017). The road network and PT system in Dalian City serve millions of trips per day, which leads to severe traffic congestion during peak hours.

There are several different PT modes in Dalian, including subways, rapid railways, and buses. The distribution of railway transportation in the city is shown in Fig. 2.1. Two subway lines (Line No. 1 and Line No. 2) primarily connect the functional areas of the city center, whereas the other rapid railway lines (Line No. 3 and Line No. 12) mainly provide longer-distance travel services for citizens.

In March 2014, the first CB route was launched in several highly demanded travel corridors in Dalian city, initiated by the state-owned PT company. After that, the CB services spread to more areas thanks to the better level of service (Wang et al. 2019). More CB routes were subsequently made available for reservation, and a few private companies collaborated with the government to provide more CB routes for citizens. The customer can get access to the service by subscribing to the preferred route using a smartphone application or an easy-to-access website.

The CB system in Dalian works as follows. First, prospective CB users are required to register on a website or online platform and submit a travel request that includes the travel OD, departure time, and desired arrival time (Liu and Ceder 2015). Upon receiving the travel requests, the CB service operators design preliminary CB routes based on the potential demand, and these routes are then uploaded onto the website or online platform to recruit additional users. After negotiating with the users regarding pick-up/drop-off times and pick-up/drop-off stations and optimizing the planned routes to reduce passenger travel time and the operation cost for the CB service

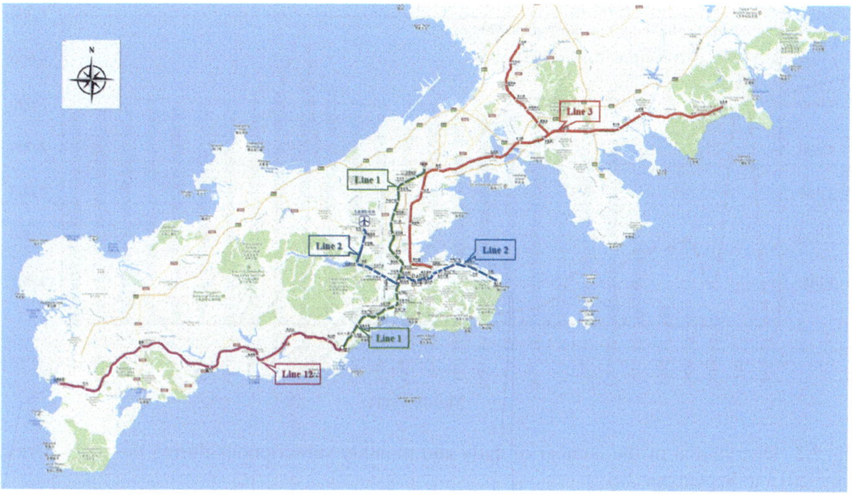

Fig. 2.1 Railway transportation network in Dalian

provider, a number of highly occupied CB routes are scheduled, while the rest are canceled.

To purchase a seat, users can either choose to subscribe to a monthly service (for one-way/round-trip regular routes on weekdays during the month) or daily service (for one-way/round-trip regular routes on a specific day) using convenient smartphone applications. The passengers are required to arrive at the appointed station by the appointed boarding time and show an electric ticket when boarding the bus. Basically, the CB routes for monthly subscribers are modified monthly according to the subscriber replacement pattern.

The number of subscriptions during the study period, from June 2015 to September 2017, is illustrated in Fig. 2.2. In the beginning, the daily subscription is more popular for more customers to start trying this new bus service instead of regarding it as a regular choice of travel directly (such as commuting). It is due to the discount coupons or lucky free tickets (the operator randomly distributed free tickets of one day CB trips to a limited number of users for experiential marketing) given to new CB service subscribers. The second peak in December 2016 arose from the strategy implemented to attract more users working in corporations in Dalian City (contrasts between the CB service provider and those corporations to provide commuting shuttle bus service for white-collars). Indeed, both of these strategies attracted a significant number of passengers, even though temporarily.

The monthly subscription became officially available in August 2015 for the first time. In December 2015, the number of monthly subscriptions first reached more than 5000 and then remained relatively steady. A great number of new monthly subscribers were generated in December 2015 and contributed the most in terms of the number of users retained. Subsequently, the retained users turned out to be the main component of passengers, and only a small number of new users generated

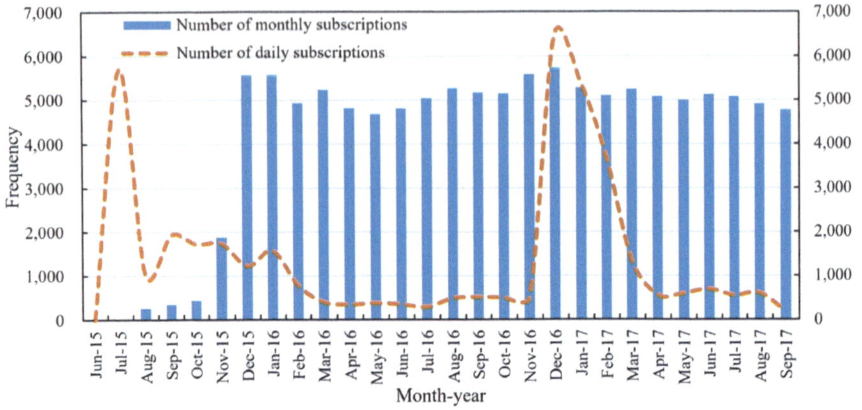

Fig. 2.2 Distribution of the number of daily and monthly subscriptions during the study period (June 2015 to September 2017)

(as illustrated in Fig. 2.3). The monthly subscribers rapidly dominated the passenger groups after the service was made officially available (Wang et al. 2019). Therefore, the subscription behavior of monthly ticket users has a greater impact on the demand evolution of the CB system.

To make the launched routes more vital and sustainable, and thus maximize passenger retention, the operator designed the policy for route launching and canceling. This policy helped to recognize the best routes in operation with enough travel demand, while in the experiment, the operator endeavored to make the policy more flexible. The CB routes with relatively lower seat occupancy rates were partially adjusted, combined, or canceled if the occupancy rate in the operation was less than 50% for several months. Therefore, the cases of route cancellation several months after the launch were quite rare.

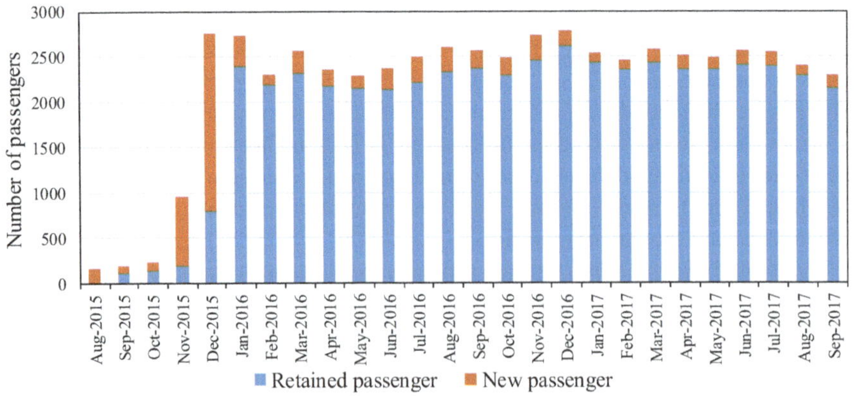

Fig. 2.3 Changes in the number of monthly subscribers during the research period

In this study, actual subscription records for the CB system in Dalian were generously provided by the service provider, which plays a vital role in initiating and operating CB services in Dalian. A total of 155,902 subscriptions were observed within this period, comprising 116,050 monthly subscriptions and 39,852 daily subscriptions. It should be noted that 87,456 out of 155,902 subscriptions are round-trip tickets whereas the others are one-way tickets.

The subscription data record details for each subscription, including corresponding coordinates of boarding/deboarding stations and the designed departure/arrival time of each route. There is a total of 11,556 CB service users, of which 69% are females and only 29% are males. The remaining 2% did not provide their gender information because of privacy concerns.

Figure 2.4 illustrates the heat maps on distributions of origins and destinations of monthly subscriptions in the morning and afternoon respectively. Major service areas of the current CB system are distributed in central and eastern Dalian where urban centers aggregate. The quite similar distribution of origins in the morning versus destinations in the afternoon and destinations in the morning versus origins in the afternoon present evidence that commuters are the main users of the monthly subscriptions. The subscriptions in the afternoon are less than those in the morning. This may be observed because the pre-reserved transit service is susceptible to the uncertainty (irregularity) of the off-duty time.

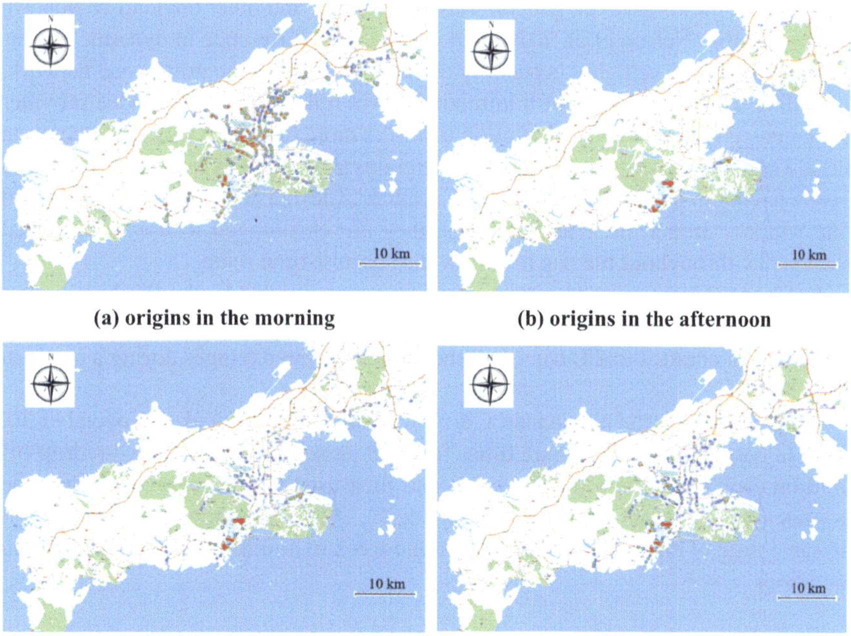

(a) origins in the morning **(b) origins in the afternoon**

(c) destinations in the morning **(d) destinations in the afternoon**

Fig. 2.4 Spatial distribution of monthly subscriptions

2.3.5 Data Collection

Based on the actual travel demand data, the routes for each OD pair were re-planned for these travel modes and the corresponding data were collected. These tasks were accomplished using the Baidu Map API, which is a service provided by Baidu Inc. that enables developers to interact directly with the Baidu Map servers.

A person who commutes regularly by driving is assumed to be familiar with the routes that he/she has driven frequently. Thus, there is a high likelihood that the driver will choose the shortest-distance path to commute to/from work without using any navigational instruments to minimize fuel consumption. In contrast, taxi drivers are primarily concerned with travel time. Limited by the working hours per day, taxi drivers aim to minimize the travel time for each customer to increase turnover. Thus, taxi drivers always opt for the shortest-time route, and the fare is charged from the passenger based on the taxi-tariff system in Dalian. For bicycling, there are few parts of roads designed with cycling paths and steep slopes can be seen on many roads. The sharing-bike is also taken as an alternative in the transportation system of Dalian. For those cyclists, the shortest path is the shortest-time route and the time-based tariff of the bike-sharing system prompts cyclists to choose the shortest-path (Lin et al. 2013).

Because the CB transit mode was not available on the Baidu Map, CB routes were generated by running the program and setting the travel mode as 'private car'. In particular, minibusses are generally used in the CB and other demand-responsive transit systems (Nelson et al. 2010) and present little difference in dynamic performance with private vehicles, especially in the peak hours of the urban road network. According to the principles of minimizing travel time, which the service provider implemented as a strategy of operation to attract more users, the shortest-time route was always chosen between every two contiguous stops of a CB route. Due to the CB drivers having to pick up and drop off passengers at designated stops within a limited time window around their arrival times, they are encouraged to use navigational instruments throughout the trip to follow the shortest-time route.

As for the PT, the least-transfer-time route recommended by the Baidu Map was selected as a representative of a PT trip, to avoid the most disadvantageous aspect of PT trips in contrast to CB trips, i.e., the number of interchanges during a trip (Liu et al. 2016).

The program for the re-planned CB routes was run on Thursday, November 30, 2017. In particular, the departure times for each passenger at each pick-up/drop-off point on each route were obtained first and then used for other travel modes (the programs were run on Thursday, December 14, 2017). The programs were run at the same time of day and the same day of the week to minimize variations of traffic conditions.

2.4 Results and Discussions

2.4.1 Descriptive Statistics of Performance Metrics

The routes for a total of 72,496 monthly subscriptions and 38,486 daily subscriptions were generated successfully with the route planning program. The descriptive information in terms of the performance metrics (expressed in terms of the mean and standard deviation) selected are summarized in Table 2.2.

Traveling by taxi consumes the least time (*TT*) and is the fastest out of the five modes (largest *ATS*). It is also the most expensive one. Bicycling consumes more time (*TT*) and the costs are quite cheap, second only to that of PT. It is beneficial for health but highly influenced by the weather conditions and topography of the road network. In the modern transportation system, the PT is always the cheapest choice

Table 2.2 Descriptive statistics of performance metrics for the five travel modes (standard deviations in parentheses follow means)

	Performance metrics				
	TT (min)	*ATS* (km/h)	*NOS*	*DAT* (min)	*TE* (RMB)
	Monthly subscriptions				
Bicycles	77	10.4	–	–	3
	(49)	(0.01)			(1.7)
CBs	52	22.69	7	7	8.9
	(31)	(6.23)	(3)	(30)	(5.3)
PT	80	11.6	16	−23	2.6
	(27)	(4.29)	(9)	(26)	(2.1)
Taxis	20	40.41	–	–	33
	(10)	(6.61)			(22.1)
Private cars	38	20.8	–	–	26.5
	(16)	(5.67)			(4.0)
	Daily subscriptions				
Bicycles	52	10.41	–	–	2.2
	(44)	(0.01)			(1.4)
CBs	40	24.37	5	12	10.1
	(36)	(6.79)	(3)	(63)	(5.2)
PT	54	10.4	13	−2	1.3
	(31)	(4.29)	(9)	(58)	(1.7)
Taxis	14	35.72	–	–	23.6
	(10)	(8.61)			(16.7)
Private cars	26	23.44	–	–	24.4
	(21)	(6.14)			(3.6)

but also accompanied with a longer *TT* and low *ATS*. The private cars provide the highest level of comfort for trips, but it is also an expensive option, second only to taxis.

From the descriptive statistics of the metrics, traveling by CBs consumes much longer *TT* than taxis and private cars and at the same time is much cheaper. In contrast to PT, traveling by CBs is faster and with much lesser passing-by stations, but is more expensive.

2.4.2 Performance Evaluation

To determine the impacts of these metrics on the utility function of these metrics and ascertain the trade-offs among them, a multinomial logit model is employed. Conventionally, multinomial logit models cannot be applied to samples such as ours, in which all cases chose the same alternative (CB). However, the directions for these metrics point to and their trade-offs, for instance, between the travel time and travel expense, are worthy to be estimated using multinomial logit models excluding alternative-specific constants.

Particularly, the alternatives are treated as unlabeled, and only generic coefficients are included in the multinomial logit model. Unlabeled alternatives are often used for route choice analysis or stated preference survey where each alternative is represented only by the level of service attributes and no label is given to each alternative such as route 1, route 2, etc. Thus, it is not necessary to specify alternative-specific constants in the unlabeled model (Sarrias and Daziano 2017), and many software packages provide options of excluding alternative-specific constant for multinomial logit model specification, such as "mlogit" in R (Croissant 2011).

There are studies indicate that the labeled or unlabeled treatment of alternative in the stated preference survey does not statistically affect the estimates of the coefficient parameters (Jin et al. 2017), and the unlabeled discrete choice experiment is more suitable to investigate the trade-offs between attributes (de Bekker-Grob et al. 2010). In our case, even though the treatment of the alternatives as unlabeled is not ideal since each mode may have alternative specific attributes to affect the mode choice, the estimates of the metrics are still meaningful to show their contributions to the utility function and trade-offs among them by roughly treating the alternatives as unlabeled.

As discussed in Sect. 3.2, the monthly subscription retains a relatively stable number of passengers and contributes in large part to the successful operation of the CB system, and therefore is focused on in this chapter. In particular, all the 72,496 monthly subscriptions (whose routes were successfully generated) were subscribed by 6912 users during the research period. For each subscriber, the most frequently used CB routes were extracted as their representative CB routes.

All five metrics were examined in multinomial logit models. Particularly, all metrics were assumed to have the same coefficients for the five modes (as generic variables), and the alternative-specific constants were excluded. To avoid multicollinearity, the *TT* was examined in the first model, while the *ATS* and *NOS* were examined in the second model. The *DAT* was transformed into two separate variables, *DAT*1 and *DAT*2. The *DAT*1 indicates the positive *DAT* (arrival at the destination earlier than expected), while *DAT*2 indicates the negative *DAT* (arrival at the destination later than expected). A summary of the estimation results is listed in.

As intuited, travelers are willing to choose a mode with shorter *TT* and lower *TE*. Particularly, the value of time is 3.8 RMB per 10 min. This indicates that travelers are willing to pay 3.8 RMB more to save 10 min on their trips, which is a little larger than the time value (3.05 RMB) transformed from the hourly minimum wage rate in Dalian in 2019 (Wikipedia Contributors 2020). Travelers prefer a mode with higher travel speed, and the *NOS* indicates negative to the utility even though it is not significant enough. The estimates of *DATs* show that arrival earlier is preferred to the later.

To understand the relative advantages of CB services, this study tested the significance of the mean differences in the metrics between CBs and PT/private cars/taxis using paired sample t-test (Zimmerman 1997). In particular, the directions that those metrics point to in Table 2.3 are the ways of performance improvement. The extent in service quality that the CB improves from PT and the differences of its performance from private cars or taxi services are also compared to reveal the actual position (role) of CB service (in terms of the performance metrics) in transportation system (how much it improves the service quality from PT and how far it is from taxis and private cars). The results are summarized in Table 2.4.

Table 2.3 Results of the multinomial logit models

	Model 1	Model 2
	Coeff. (z-value.)	Coeff. (z-value.)
TT (10 min)	−0.435*	
	(−19.69)	
TE (RMB)	−0.113*	−0.161*
	(−25.06)	(−25.15)
ATS (km/h)		0.125*
		(26.92)
NOS		−0.006
		(−1.36)
DAT1 (min)	0.442*	0.447*
	(33.82)	(33.73)
DAT2 (min)	−0.010*	−0.001
	(−6.57)	(−0.68)
log-likelihood	−3460	−3208

Note '*' indicates that $p < 0.01$

Table 2.4 Results of paired mean difference t-tests

Variable	Mean	95% CI	p-value
$TT_{CB} - TT_{PT}$	−23.07 min	−23.24– −22.91	<0.001
$ATS_{CB} - ATS_{PT}$	12.09 km/h	12.05– 12.12	<0.001
$NOS_{CB} - NOS_{PT}$	−8.58	−8.63– −8.53	<0.001
$DAT_{CB} - DAT_{PT}$	24.62 min	24.43–24.81	<0.001
$TE_{CB} - TE_{CAR}$	−16.48 RMB	−16.50– −16.45	<0.001
$TE_{CB} - TE_{TAXI}$	−20.44 RMB	−20.55– −20.33	<0.001
Null hypothesis: the travel time savings in shifting from PT to CBs and from CBs to private cars do not have a significant difference			
$(TT_{PT} - TT_{CB}) - (TT_{CB} - TT_{CAR})$	9.03 min	8.74–9.32	<0.001
Null hypothesis: the travel cost increases in shifting from PT to CBs and from CBs to private cars do not have a significant difference			
$(TE_{CB} - TE_{PT}) - (TE_{CAR} - TE_{CB})$	−10.02 RMB	−10.08– −9.96	<0.001
Null hypothesis: the travel cost increases in shifting from PT to CBs and from CBs to taxis do not have a significant difference			
$(TE_{CB} - TE_{PT}) - (TE_{TAXI} - TE_{CB})$	−12.03 RMB	−12.14– −11.93	<0.001

Note All the subscriptions (including monthly and daily subscriptions) were included ($N = 110{,}982$)

Compared to PT, the performances of CB service are significantly improved in terms of *TT*, *ATS*, *NOS*, and *DAT*, even with an amount of increase in expense. Particularly, the cost of a trip by CB is still significantly far from that of a taxi (20.44 RMB) or private car (16.48 RMB).

The results in terms of the mean difference of *TT* between CB versus PT and CB versus private cars further suggests that the time saved by shifting from PT to CB is significantly larger than that of shifting from CB to a private car. The costs increased by shifting from PT to CB are significantly less than that of shifting from CB to private cars and taxis. In particular, the results illustrate that the CB service is an intermediate choice that can balance service quality and costs between traveling by PT and private cars/taxis. With a limited increase in costs from PT, the CB service significantly improves the service quality.

2.4.3 Spatial and Temporal Coverage

Based on the multinomial logit model calibrated, the probability of choosing the CB travel can be calculated for each individual. In particular, the probabilities in each zone are averaged and then multiplied by the population to indicate the advantages of traveling by CB (on the township level). The analysis of the relationship between the spatial distribution of the advantage and actual demand will help reveal the mechanism of the spatial distribution in demand. The population data is sourced

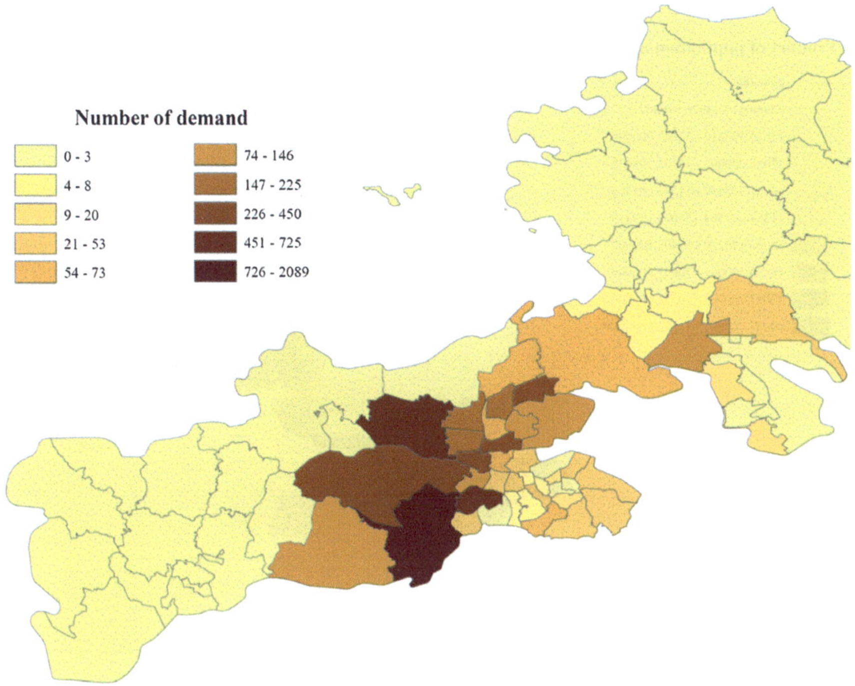

Fig. 2.5 Distribution of the actual demand of the CB system on the township scale

from the population census of China conducted in 2010 on the township scale (Feng 2012), which is the latest population census conducted in China.

Figure 2.5 shows the spatial distribution of the actual demand. Most of the demand is concentrated in the central area of Dalian, as shown by the darker-colored areas. There is no demand served in the most western and northern areas of the city.

Figure 2.6 illustrates the spatial distribution of the product of population and mean probability on the township scale, indicating the relative advantage of subscribing to a CB service on the zone level (township scale).

The relationship between the relative advantage of using a CB service and the actual demand is examined by regressing the actual demand with respect to the relative advantage. The results of the regression models are listed in Table 2.5. The results show a significantly positive relationship between the relative advantage and the demand. However, the residuals of the ordinary least squares model exhibit spatial clustering, indicating that the relationship between the explanatory variables and the dependent variable are spatially auto-correlated. After considering the spatial auto-correlation using a geographically weighted regression model (Fotheringham et al. 1996), more than 64% of the variation in the demand can be explained by the relative advantage by zone.

Comparing Figs. 2.5 and 2.6 reveals that most zones are covered with the same or similar colors, particularly the high-demand zones in central Dalian. Nevertheless,

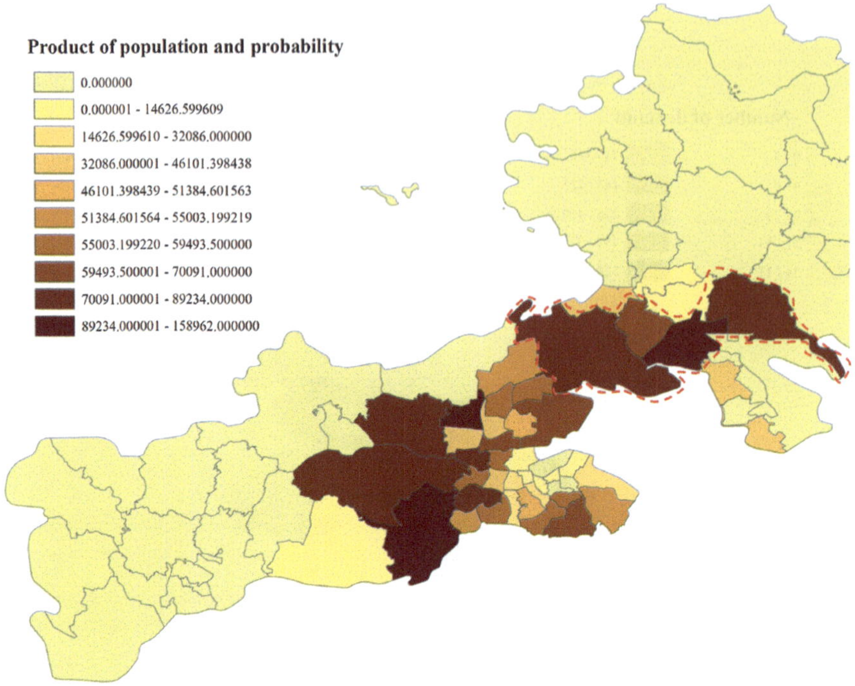

Fig. 2.6 Distribution of the product of the population and mean probability on the township scale

Table 2.5 Estimation results of the ordinary least squares regression and geographically weighted regression models

	Ordinary least squares regression model		
Variable	Coefficient	Std. Err	t-value
Intercept	−27.5257	31.5135	−0.8735
Relative advantage	0.0041**	0.0007	5.8036
N	80		
R-squared	0.302		
AICc	1094		
	Geographically weighted regression model		
Bandwidth	0.148		
Sigma	169.310		
R-squared	0.642		
AICc	1058		

Note AICc denotes Akaike's Information Criterion; '**' indicates that $p < 0.001$

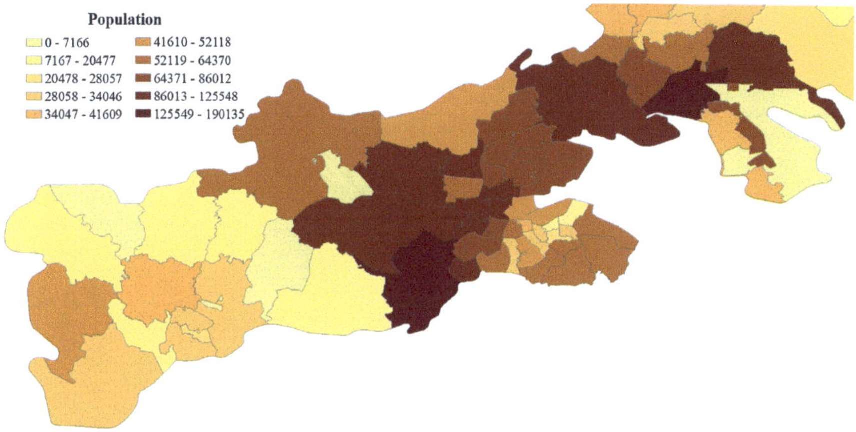

Fig. 2.7 Distribution of population on the township scale

a big difference is shown in some areas in eastern Dalian, particularly in the zones surrounded by the red dotted line in Fig. 2.6. It is a newly developed city subcenter with an increasing population in eastern Dalian. From the distribution of population illustrated in Fig. 2.7, we can see the great number of residents collected in these areas which naturally lead to a larger amount of demand to use CB service. The advantages of traveling by CB in these areas (surrounded by the red dotted line) show consistent higher values among areas of eastern Dalian.

The destinations of the demand in these areas are concentrated in the several zones where the city centers are located in the central Dalian, as illustrated in Fig. 2.8. However, the CB demand flows from these areas, obviously, experience longer distances than the demand in the central Dalian. Undoubtedly, the bus sharing system is facing more challenges with long-distance routes, as it would have more difficulties in finding enough partners due to the small number of trips from the same origin zone to the same destination zone with similar departure/arrival times requested (Correia and Viegas 2009) and make the demand-responsive transit service unreliable (Mageean and Nelson 2003). It is not strange that only a few routes were successfully launched to serve demand from these areas to the central Dalian, therefore, leading to the big difference between the theoretically estimated and the actual demand. On the other hand, the difference suggests a marketing potential for a more flexible and complicated route design of CB service, for instance, once-transfer-permitted route design, especially for long-distance travels.

Based on the demand from online surveys, the pick-up and drop-off stations for each route are relatively concentrated. The departure time and arrival time of each passenger dispatched in a single bus are scheduled to be as similar as possible. The temporal coverage of monthly CB subscriptions is illustrated in Fig. 2.9. The results show that the departure times of the demand are concentrated from 07:00 to 08:00 in the morning and from 17:30 to 18:30 in the afternoon. The arrival times are concentrated from 08:00 to 09:00 in the morning and from 18:30 to 19:30 in the

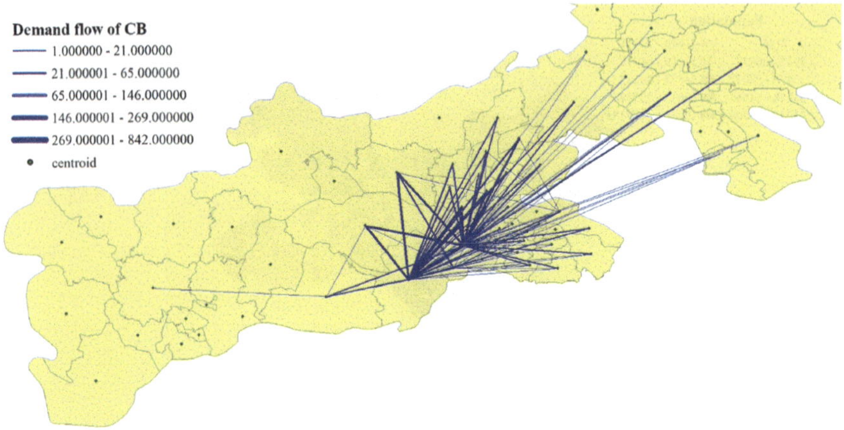

Fig. 2.8 Desire line of CB demand

afternoon. As a result, the time period during travel falls exactly in the rush period of the road network in Dalian. Thus, the reductions in travel time for passengers who shift from PT to CB and in traffic load by attracting private car drivers to CB contribute to the improvement of traffic conditions, especially during peak hours.

To investigate the spatial coverage of the CB service and its relationship to PT services, the number of PT stations around each CB demand point is counted. In particular, information on the PT stations in the whole city area, including bus, subway, and rapid railway stations, were collected by mining the points of interest in Gaode Map on May 4, 2018. The resulting data mainly includes the coordinates for each station. The number of PT stations was counted within a buffer of 500 m for each CB demand point. The results are illustrated in Fig. 2.10. For 6912 demand points (for the monthly subscribers), more than 3% (i.e., more than 200 users) are located in areas without any PT stations within 500 m. In these cases, the CB system provides an extra mode choice and serves as the main bus transit service. In more than 73% of the cases, the number of PT stations within 500 m is less than four, suggesting that the CB service primarily acts to supplement the PT service in areas with poor coverage levels.

2.5 Conclusions

After failures were experienced in the development of CB systems in some cities, the feasibility of this new bus sharing system is being questioned. Blindly developing a CB system before being clear regarding its role in the overall transportation realm can lead to a huge waste of resources in those cities that experience a drastic decline in customers. Thus, this study investigates the actual role of CB services in the overall transportation realm based on a successful case, as the role of the CB service

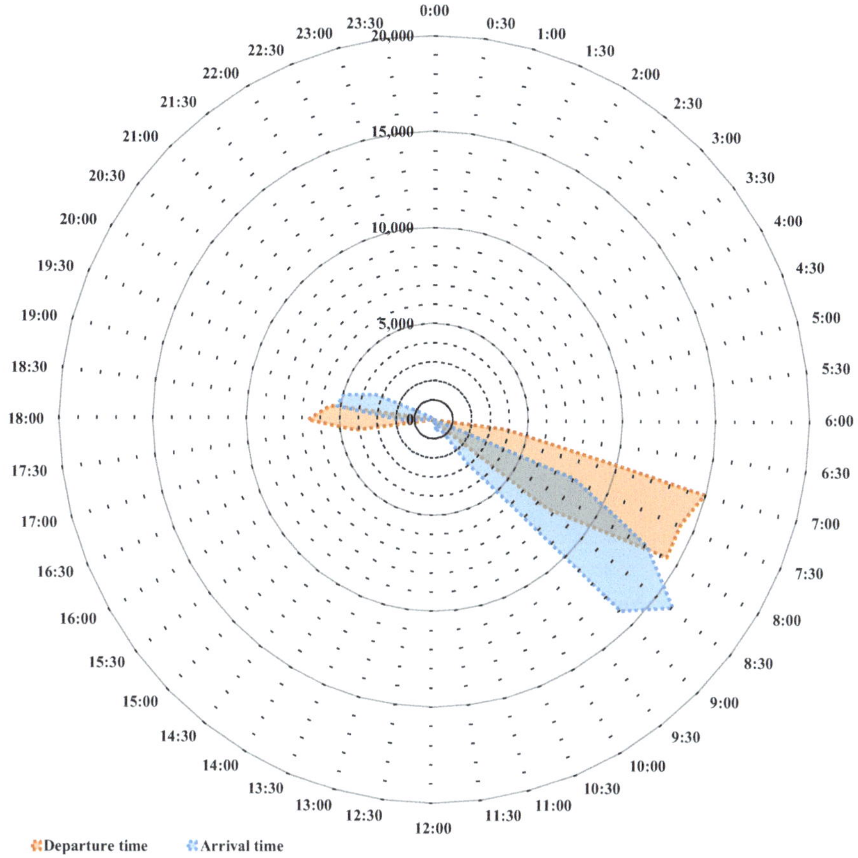

Fig. 2.9 Distribution of the departure and arrival times for monthly subscriptions

determines its primary service object, system construction, marketing orientation, and even government function.

The directions on the utility function of these metrics were determined using a multinomial logit model excluding alternative-specific constant. The results illustrate that the CB service is an eclectic choice that can balance service quality and costs between traveling by PT and private cars/taxis. Comparison of performance metrics for all available transportation modes shows that the CB service significantly improves the service quality (in terms of the travel time, travel speed, number of stations, and differential arrival time) compared to traditional PT services, even though with an increased travel expense. In particular, the costs increased by shifting from PT to CB are significantly less than that of shifting from CB to private cars and taxis.

The temporal coverage reveals that the period of peak travel with the CB services falls exactly in the rush period of the road network in Dalian. Thus, the reductions in

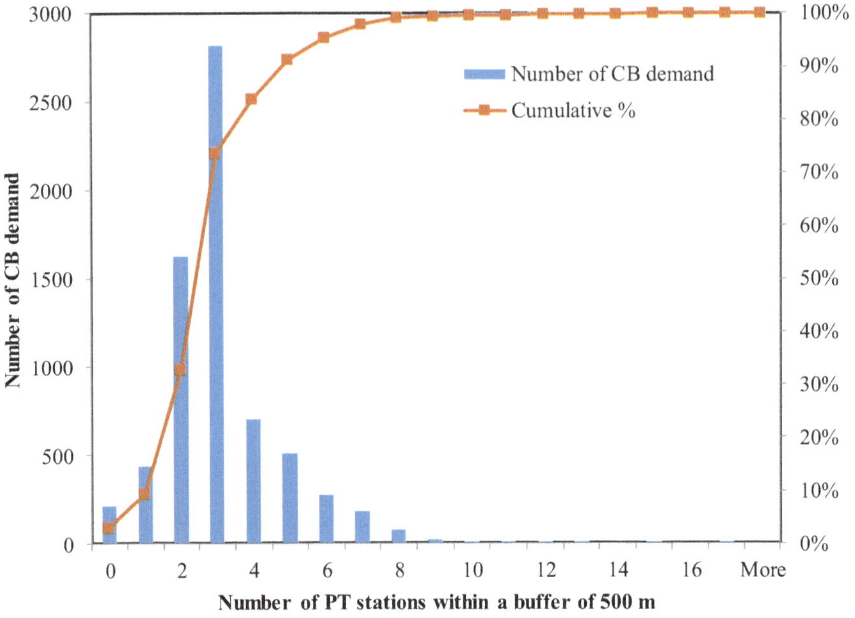

Fig. 2.10 Number of PT stations within 500 m of each CB demand point

travel time for passengers who shift from PT to CB and in traffic load achieved by attracting private car drivers are contributing to the alleviation of traffic congestions during rush hours. The CB service mainly acts to complement traditional PT services in areas with poor coverage levels of PT services, while it serves as the main bus transit service in several areas (more than 200 subscribers) without PT service within 500 m.

References

Chowdhury, S., A. Ceder (Avi), and B. Schwalger. 2015. The effects of travel time and cost savings on commuters' decision to travel on public transport routes involving transfers. *Journal of Transport Geography* 43: 151–159. https://doi.org/10.1016/j.jtrangeo.2015.01.009.

Correia, G., and J.M. Viegas. 2009. A conceptual model for carpooling systems simulation. *Journal of Simulation* 3: 61–68. https://doi.org/10.1057/jos.2008.4.

Croissant, Y. 2011. Estimation of multinomial logit models in R: The mlogit Packages. *Journal of Statistical Software* 95. https://doi.org/10.18637/jss.v095.i11

Dalian Municipal Bureau of Statistics. 2017. Dalian Statistic Yearbook 2017 [WWW Document]. URL http://www.stats.dl.gov.cn/index.php?m=content&c=index&a=lists&catid=57. Accessed 31 Mar 2018.

Davison, L., M. Enoch, T. Ryley, M. Quddus, and C. Wang. 2014. A survey of demand responsive transport in great Britain. *Transport Policy (Oxf)* 31: 47–54. https://doi.org/10.1016/j.tranpol.2013.11.004.

de Bekker-Grob, Esther, W., L. Hol, B. Donkers, L. Van Dam, J.D.F. Habbema, M.E. Van Leerdam, E.J. Kuipers, M.-L. Essink-Bot, and E.W. Steyerberg. 2010. Labeled versus unlabeled discrete choice experiments in health economics: An application to colorectal cancer screening. *Value in Health* 13: 315–323

de Oña, R., J. de Abreu e Silva, C. Muñoz-Monge, and J. de Oña. 2018. Users' satisfaction evolution of a metropolitan transit system in a context of economic downturn. *International Journal of Sustainable Transportation* 12: 66–74. https://doi.org/10.1080/15568318.2017.1328546

Dell'Olio, L., A. Ibeas, and P. Cecin. 2011. The quality of service desired by public transport users. *Transport Policy (Oxf)* 18: 217–227. https://doi.org/10.1016/j.tranpol.2010.08.005.

Eboli, L., and G. Mazzulla. 2011. A methodology for evaluating transit service quality based on subjective and objective measures from the passenger's point of view. *Transport Policy (Oxf)* 18: 172–181. https://doi.org/10.1016/j.tranpol.2010.07.007.

Feng, N. 2012. *Tabulation on the population census of the people's republic of China by township.* Beijing: China Statistics Press.

Fotheringham, A.S., C. Brunsdon, and M.E. Charlton. 1996. Geographically weighted regression: A method for exploring spatial nonstationarity. *Geographical Analysis* 28: 281–298. https://doi.org/10.1111/j.1538-4632.1996.tb00936.x.

Jin, W., H. Jiang, Y. Liu, and E. Klampfl. 2017. Do labeled versus unlabeled treatments of alternatives' names influence stated choice outputs? Results from a mode choice study. *PLoS ONE* 12: 1–24. https://doi.org/10.1371/journal.pone.0178826.

Jintounet. 2017. Real time gasoline price inquiry [WWW Document]. URL http://www.cngold.org/c/2017-12-14/c5541781.html. Accessed 31 Mar 2018.

Kirby, R.F., and K.U. Bhatt. 1975. An analysis of subscription bus experience. *Traffic Quarterly* 29: 403–425.

Li, J., X. Chen, X. Li, and X. Guo. 2013. Evaluation of public transportation operation based on data envelopment analysis. *Procedia—Social and Behavioral Sciences*: 148–155. https://doi.org/10.1016/j.sbspro.2013.08.020.

Lin, J.R., T.H. Yang, and Y.C. Chang. 2013. A hub location inventory model for bicycle sharing system design: Formulation and solution. *Computers & Industrial Engineering* 65: 77–86. https://doi.org/10.1016/j.cie.2011.12.006.

Liu, T., and A. Ceder. 2015. Analysis of a new public-transport-service concept: Customized bus in China. *Transport Policy (Oxf)* 39: 63–76. https://doi.org/10.1016/j.tranpol.2015.02.004.

Liu, T., A.A. Cedera, R. Bologna, and B. Cabantous. 2016. Commuting by customized bus: A comparative analysis with private car and conventional public transport in two cities. *Journal of Public Transportation* 19: 55–74. https://doi.org/10.5038/2375-0901.19.2.4.

Mageean, J., and J.D. Nelson. 2003. The evaluation of demand responsive transport services in Europe. *Journal of Transport Geography* 11: 255–270. https://doi.org/10.1016/S0966-6923(03)00026-7.

Ministry of Industry and Information Technology of the People's Republic of China. 2015. Interpretation of the fourth stage standard of passenger car fuel consumption [WWW Document]. URL http://www.miit.gov.cn/n1146285/n1146352/n3054355/n3057585/n3057589/c3616982/content.html. Accessed 31 Mar 2018.

National Bureau of Statistics of the People's Republic of China. 2011. Tabulation on the 2010 population census of the People's Republic of China [WWW Document]. URL http://www.stats.gov.cn/tjsj/pcsj/rkpc/6rp/indexch.htm. accessed 31 Mar 2018.

Nelson, J.D., and T. Phonphitakchai. 2012. An evaluation of the user characteristics of an open access DRT service. *Research in Transportation Economics* 34: 54–65. https://doi.org/10.1016/j.retrec.2011.12.008.

Nelson, J.D., S. Wright, B. Masson, G. Ambrosino, and A. Naniopoulos. 2010. Recent developments in flexible transport services. *Research in Transportation Economics* 29: 243–248. https://doi.org/10.1016/j.retrec.2010.07.030.

Price Bureau of Dalian City. 2014. Notice on printing and distributing charging standard of motorized parking service [WWW Document]. URL http://www.wjj.dl.gov.cn/htmlfile/3/21557.html?xyz=0.3004350671436528. Accessed 31 Mar 2018.

Redman, L., M. Friman, T. Gärling, and T. Hartig. 2013. Quality attributes of public transport that attract car users: A research review. *Transport Policy (Oxf)* 25: 119–127. https://doi.org/10.1016/j.tranpol.2012.11.005.

Sarrias, M., and R. Daziano. 2017. Multinomial logit models with continuous and discrete individual heterogeneity in R: The gmnl package. *Journal of Statistical Software* 79: 1–46. https://doi.org/10.18637/jss.v079.i02

Sun, C., X. Chen, H.M. Zhang, and Z. Huang. 2018. An evaluation method of urban public transport facilities resource supply based on accessibility. Journal of Advanced Transportation 2018. https://doi.org/10.1155/2018/3754205.

Wang, J., T. Yamamoto, and K. Liu. 2019. Role of customized bus services in the transportation system: Insight from actual performance. *Journal of Advanced Transportation* 2019: 6171532. https://doi.org/10.1155/2019/6171532.

Wikipedia Contributors. 2020. List of minimum wages by country [WWW Document]. Wikipedia. URL https://en.wikipedia.org/w/index.php?title=List_of_minimum_wages_by_country&oldid=952957019. Accessed 24 Apr 2020.

Zimmerman, D.W. 1997. Teacher's corner: A note on interpretation of the paired-samples t test. *Journal of Educational and Behavioral Statistics* 22: 349–360.

Zou, L., H. Dai, E. Yao, T. Jiang, and H. Guo. 2014. Research on assessment methods for urban public transport development in China. Computational Intelligence Neuroscience 2014. https://doi.org/10.1155/2014/941347.

Chapter 3
Potential Benefits of Customized Bus Services in Commuting Travel

Abstract Commuting is a major component in the creation of traffic and travel problems. Thus, more attention should be given to its practice. Private car (PC) transport, the dominant mode of commuting in most of the world's major cities, creates traffic-related social problems such as traffic congestion, traffic fatalities and injuries, and adverse environmental impacts. This study proposes a novel commuting travel mode—a customized bus (CB) transit system that provides advanced, personalized, and flexible demand-interactive minibus service using Internet, telephone, and smartphone apps. The aim was to assess and compare the performance of CB with PC and with conventional public transport (PT) systems. A methodological analysis framework was constructed to quantify operational performance measures that enable the comparison of the different travel modes. This analysis framework was then applied to two cities—Auckland, New Zealand, and Paris, France—to assess the overall performance metrics of PC, PT, and CB, such as travel costs, travel time, and fuel consumption. This comparison sheds light on the differences between the travel modes, their viability, and their competitiveness. The results of the case study show that PC is the fastest commuting mode, but the travel costs incurred by it are twice as much as for PT and CB and involve higher fuel consumption. CB also can provide a useful alternative for commuter trips in Auckland and Paris. For increased commuter trips, CB proved to be more efficient than the PC and PT modes. Finally, the CB system tends to be more profitable in Auckland than in Paris.

Keywords Commuting · Customized bus · Private car · Public transport · Travel mode

3.1 Introduction

In recent years, the rapid urbanization and increasing population density in metropolitan areas have led to significant challenges in urban mobility. Traditional public transport systems, while efficient, often fail to meet the dynamic and individualized needs of travelers, especially commuters. Conventional public transport

J. Wang et al., *Customized Bus: User Travel Behavior and Demand Evolution*, Urban Sustainability, https://doi.org/10.1007/978-981-96-3488-0_3

(PT) systems, especially those involving buses, use an old concept involving fixed routes, fixed stops, fixed terminals, fixed timetables, fixed vehicle scheduling, and driver scheduling. This traditional PT concept produces services that are not always appealing and do not necessarily attract commuters (Ceder, 2021). Congestion, long waiting times, and lack of direct routes have contributed to a decline in public transport usage, with many commuters preferring private vehicles despite their higher environmental and economic costs.

Commuting is a ubiquitous social activity that plays a crucial role in daily life and accounts for a significant portion of total household trips. Commuting is often perceived as a daily activity that generates the lowest level of positive affect and a relatively high level of negative affect (AASHTO 2015). Long-distance commuting, in particular, can lead to various physical, financial, and mental issues, such as stress and out-of-pocket costs. It also negatively impacts work-family balance and increases the risk of divorce (Sandow 2011; Stutzer and Frey 2008).

In most major cities globally, private car (PC) transport remains the predominant mode of commuting. Rapid economic growth and employment expansion have increased the number of commuters. This, combined with extensive urban sprawl, has escalated the use of PCs, leading to numerous traffic-related social issues, including heightened traffic congestion, increased traffic fatalities and injuries, and negative environmental impacts. Historically, the continual expansion of road networks and the implementation of demand-management measures have proven ineffective in mitigating these adverse effects. Consequently, the need for an efficient, reliable, and affordable PT system has become increasingly urgent. To transform commuting into a more pleasant experience, it is essential to develop an advanced and appealing PT system. As highlighted by Ceder, such a system should operate reliably and relatively swiftly, offering smooth, synchronized transfers as part of the door-to-door passenger journey (Ceder 2016).

The concept of Customized Bus (CB) services has emerged as a solution to the commuting issues. CB services are designed to bridge the gap between traditional fixed-route public transit and private car usage by offering flexible, demand-responsive transit solutions. The motivation for developing such services stems from the need to enhance the efficiency, convenience, coordination and sustainability of urban transport systems. By tailoring routes and schedules based on reservation demand and user preferences, CB services aim to provide a more personalized and efficient commuting experience (Liu and Ceder 2015).

CB services can adapt to the specific needs of commuters, providing more direct routes and reducing transfer times. This flexibility makes PT more attractive to users who require tailored travel options. By optimizing routes and schedules based on actual demand, CB services can reduce operational inefficiencies associated with fixed-route systems. This leads to shorter travel times and reduced congestion. For commuters, CB services can offer a cost-effective alternative to private car usage. For transit operators, optimized operations can lead to cost savings through improved resource utilization. What's more, by encouraging the use of PT over private vehicles, CB services can contribute to reducing traffic congestion and lowering greenhouse gas emissions, promoting a more sustainable urban environment.

Recent experiences have demonstrated that CB services can provide an attractive PT commuting system, offering advanced, personalized, and flexible demand-responsive services to commuters. CB has been successfully launched and implemented in numerous cities worldwide, including Beijing, Lisbon, and Singapore (Tong et al. 2017). This service holds significant potential for addressing the growing and diverse commuting needs of large populations and enhancing the overall commuting experience.

3.2 Literature Review

Commuting is an essential daily activity that significantly affects urban life. The predominant use of PCs for commuting has led to numerous challenges, including traffic congestion, increased travel times, environmental degradation, and public health concerns. Despite these issues, the convenience, privacy, and perceived reliability of private cars continue to make them the preferred choice for many commuters. Conversely, PT offers numerous societal benefits, such as reduced traffic congestion, lower environmental impact, and improved public health outcomes. However, PT systems must address several barriers to attract more commuters, including reliability, convenience, and overall user experience.

The decision to choose between private cars (PC) and public transportation (PT) is influenced by various factors, making it a complex process. These factors include trip characteristics such as purpose, timing, and regularity, as well as demographic characteristics like age, gender, and income level (Galdames et al. 2001; Tertoolen et al. 1998). By addressing these factors through innovative solutions like CB services, urban transport systems can encourage a shift towards more sustainable commuting practices. This shift is essential for mitigating the adverse effects of PC dominance and fostering a more efficient, sustainable, and inclusive urban mobility system (Ceder 2021; Liu et al. 2021).

Studies have shown that PCs are generally perceived to be more attractive than PT due to their convenience, flexibility, independence, comfort, speed, and reliability. Additionally, driving is often considered pleasurable and carries a certain social status. Research indicates that PT use tends to decline with increasing age and income. Women, however, show a slightly higher probability of using PT for non-commuting trip purposes (Kuhnimhof et al. 2006). Other significant factors influencing mode choice include the quality of PT services, connectivity issues, out-of-pocket travel costs, access distances to and from stations, and the overall distance between home and work (Galdames et al. 2001; Kingham et al. 2001).

Psychological resistance to switching from PC to PT is another critical factor. Studies have shown that PC users are often resistant to change due to the perceived benefits of PCs. These benefits are not only instrumental, such as freedom, comfort, and convenience, but also symbolic, representing social status, and affective, as driving is perceived as pleasurable (Beirão and Sarsfield Cabral 2007; Hiscock et al. 2002).

Once individuals acquire private vehicles, their use often shifts from being a luxury to a necessity, becoming a habitual choice for many travelers. The increasing complexity of trip chains, due to changing household travel patterns and more women entering the workforce, also poses a significant barrier to PT use (Anable 2005).

Given these challenges, developing strategies to encourage a mode switch from PC to PT, particularly for commuters, remains a critical area of interest for transportation specialists. Effective methods include improving the quality and connectivity of PT services, offering economic incentives, and addressing psychological barriers through targeted public awareness campaigns. These strategies aim to make PT a more attractive and viable option, thereby fostering a more sustainable urban mobility system.

By conducting a comprehensive comparative analysis of commuters' mode choices between CB, traditional PT, and PC, transportation planners and policy-makers can develop more effective, efficient, and sustainable transportation systems that better meet the needs of commuters (Chen et al. 2021; Liu et al. 2016).

3.3 Methodological Framework

Liu et al. (2016) introduced a methodological framework for conducting the comparative analysis of commuting travel modes, including CB, conventional PT and PC, to show the potential benefits of using CB as a commuting mode. As shown in Fig. 3.1, this methodological framework is constructed by an input-component output format, elaborating the systematic decision sequence and process of the analysis. The output of each component positioned higher in the sequence becomes an important input into lower-level decisions. This analysis framework is customized to achieve three objectives: (i) constructing an initial methodological framework for quantifying operational performance measures, such as travel time, travel cost, energy consumption, (ii) applying this framework to assessing the overall performance of PC, PT, and CB in case studies in Auckland and Paris, (iii) conducting comparisons between the cities and provide recommendations for actual CB service improvement and implementation.

Accordingly, in order to determine and compare the performance measures for the three different commuting modes in Auckland and Paris, we employed this four stages framework.

The first stage involves identifying the study areas and establishing a preliminary road network map along with PT networks. In the second stage, potential commuting trips within the selected study areas are identified. Data on trip origins/destinations, expected arrival times at destinations, and the type of vehicle used are collected. This demand collection process can be efficiently completed using a human–computer interactive system integrated with various information platforms, such as the Internet, telephone, and smartphone apps, with active participation from both users and operators. During this stage, grouping and vehicle routing strategies are employed to design and route CB vehicles to pick up commuters from their origins

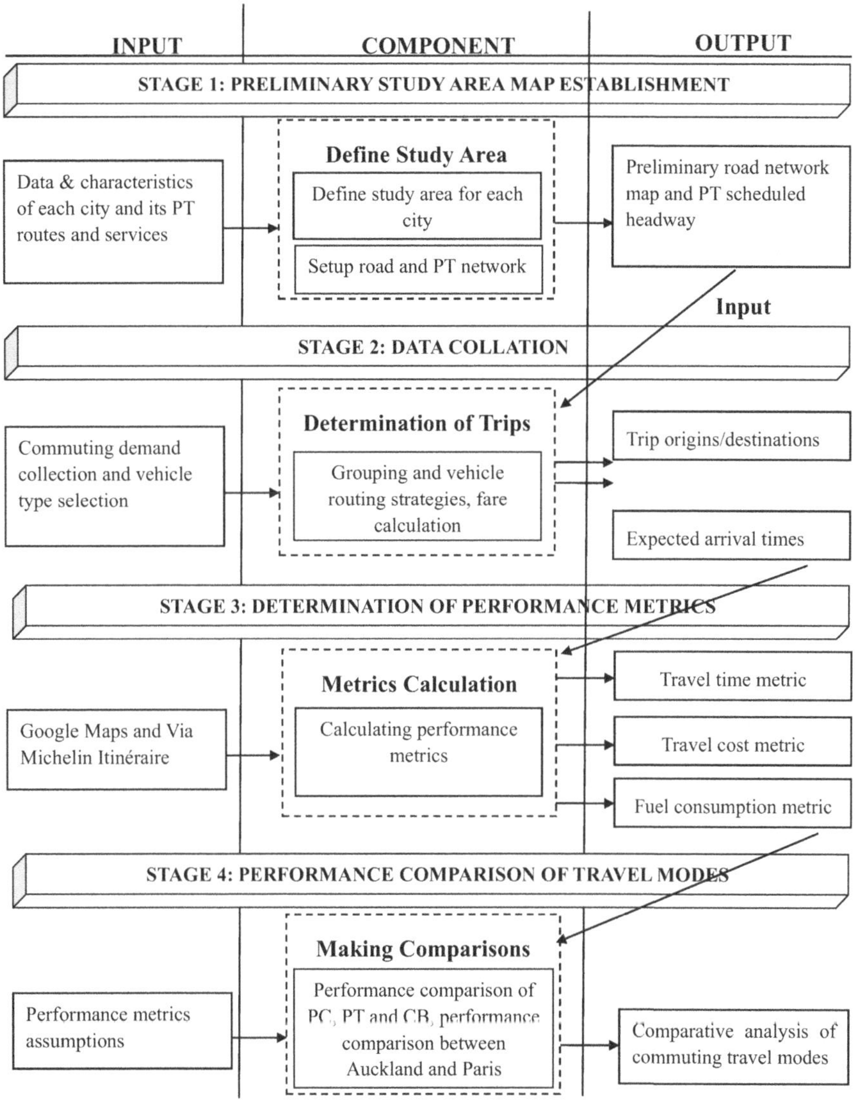

Fig. 3.1 Methodological framework for comparative analysis on commuting mode choice (Liu et al. 2016)

and drop them off at their destinations. After generating the routes, tools like Google Maps and the Via Michelin Itinéraire smartphone apps are used to collect data on trip travel time, travel cost, and fuel consumption based on simplified assumptions. In the final stage, performance comparisons between PC, PT, and CB are conducted, as well as performance comparisons between cities such as Auckland and Paris. The calculated performance metrics provide valuable policy implications and managerial insights.

Overall, the performance metric comparisons across different commuting modes and cities help us better understand the viability and competitiveness of various travel modes under different conditions. This understanding is crucial for planning future improvements to CB services. Comparative analysis helps to understand the factors influencing commuters' choices, such as convenience, cost, travel time, and comfort. By understanding commuters' preferences, CB agencies can tailor services to meet the needs of commuters better, increasing satisfaction and ridership.

3.4 Study Area and Data Collection

Liu et al. (2016) applied the methodological framework to analyze the commuting mode choice in Auckland, New Zealand, and Paris, France.

3.4.1 Study Area Description

Auckland, the largest city in New Zealand, is home to a diverse and growing population. As of the most recent estimates, the population of Auckland is approximately 1.7 million, accounting for about one-third of the entire country's population. The city is characterized by a multicultural demographic, with significant communities from the Pacific Islands, Asia, and Europe, contributing to its vibrant cultural landscape. Auckland is the economic powerhouse of New Zealand, generating around 38% of the national GDP.

Auckland's transport system is extensive, incorporating a network of roads, motorways, and PT options designed to cater to its large and dispersed population. However, the city's rapid growth has put pressure on its transport infrastructure, leading to challenges such as traffic congestion, especially during peak hours. Auckland's PT system includes buses, trains, and ferries, all integrated under the Auckland Transport (AT) network. Despite improvements and expansions, PT usage is still relatively low compared to car use, but ongoing investments aim to increase its attractiveness and efficiency. Auckland's efforts to enhance its transport system are critical to supporting its continued growth and maintaining its status as New Zealand's leading metropolis. The integration of advanced transport solutions and sustainable practices will be key to addressing the challenges posed by urbanization and ensuring the city remains an attractive place to live and work.

The case study area in Auckland is shown in Fig. 3.2, which includes the Auckland city center, the regions of Newmarket, Grey Lynn, Epsom, Penrose, Remuera, Ellerslie and Mount Wellington. All possible buses and three train lines in this area, namely the Southern line, the Eastern line and the Western line, are taken into consideration. For the comparative study, only travel between Auckland suburbs and the CDB, which are daily work trips, are considered.

Paris, the capital city of France, is one of the most populous urban areas in Europe. The Paris metropolitan area, known as Île-de-France, has a population of approximately 12 million people. The city itself has around 2.1 million residents. Paris is known for its diverse population, including a significant number of immigrants and expatriates, contributing to its rich cultural and social fabric. Paris is a global economic powerhouse, contributing significantly to the national and European economy. The GDP of the Île-de-France region is around €700 billion, making it the wealthiest region in Europe.

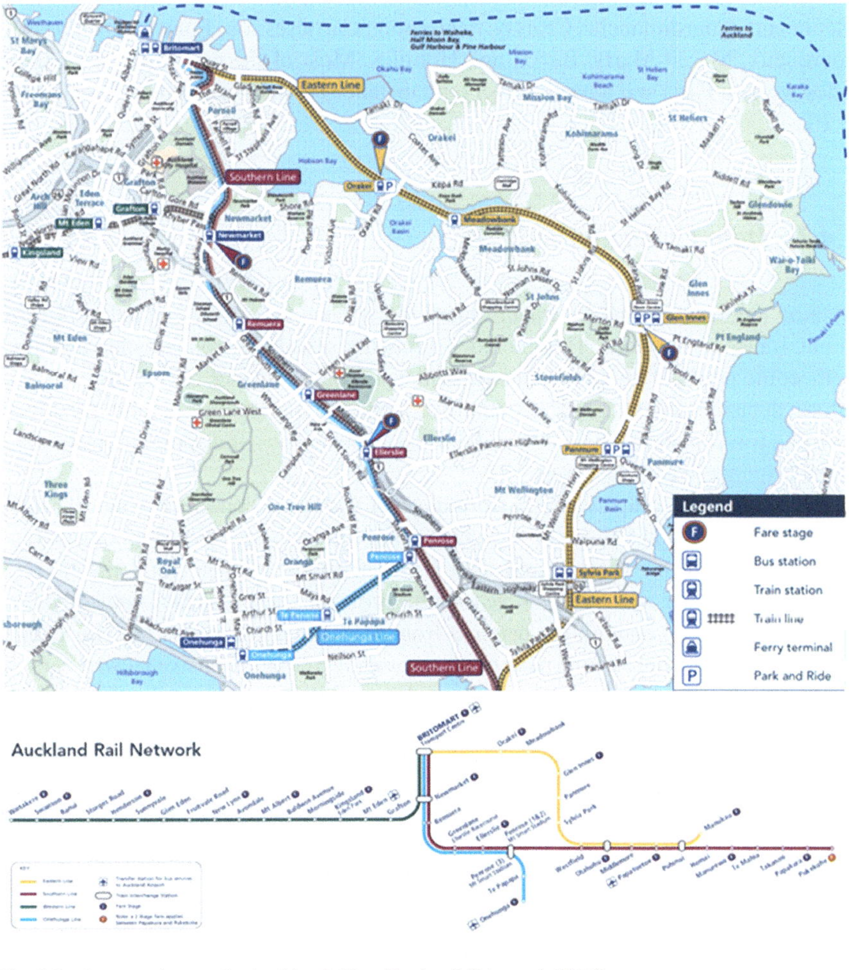

Fig. 3.2 Case study area in Auckland, New Zealand (Liu et al. 2016)

Paris boasts a sophisticated and extensive transport system designed to cater to its large population and the influx of tourists and business travelers. The city's transport infrastructure includes a comprehensive network of roads, an extensive public transport system, and several major airports. Paris has one of the most efficient and comprehensive public transport systems in the world, managed primarily by RATP (Régie Autonome des Transports Parisiens) and SNCF (Société Nationale des Chemins de fer Français). The RATP Group is the PT operator and provides a complex PT network that consists of trains, buses, metros, and tramways. It is estimated that there are annually three billion PT users. Paris continues to invest in its public transport infrastructure to meet growing demand and address sustainability goals.

The case study area in Paris is shown in Fig. 3.3, which includes La Défense and the western part of metropolitan Paris, which is composed of the cities of Vauréal, Ménucourt, Courdimanche, Croissy, Aigremont, Carrières sous Poissy, Chambourcy, Fourqueux, Mareil-Marly, Bezon and Houilles. Most of the people living in these cities work in Paris and have to travel long distances to and from work, either by PC or PT.

3.4.2 Data Collection

Data on passenger volumes, travel times, and service frequency can help optimize public transport services. This includes adjusting schedules, routes, and capacities to meet the needs of commuters better. Transport data provides critical insights that inform decisions related to service improvements, and policy formulation. By understanding travel patterns, demand, and performance metrics, transport authorities can make evidence-based decisions.

To effectively compare the performance of the three commuting modes—PC, conventional PT, and CB—four key performance metrics are identified: the average difference between expected and actual arrival times, average total travel time, average total travel cost, and average total fuel consumption. The two study areas are studied, examining commuting patterns from 6:45 A.M. to 10:00 A.M. This period encompasses both normal and peak hours, providing a comprehensive view of various traffic conditions and ensuring a robust analysis of each mode's performance.

In the two study areas, trip origins where commuters have access to both PC and PT options are considered potential service points. To ensure a representative sample, we examined 100 trips in Paris and 100 trips in Auckland. For the Paris study, we selected five different train stations and identified 20 addresses in the vicinity of each station, aiming to cover a broad area. This process was facilitated using Google Maps.

In the study of Liu et al. (2016), a smartphone app, named Via Michelin app, was employed to collect the PC travel data. The app allows users to select distance and cost parameters, as well as vehicle and fuel types. Users can also adjust fuel costs manually. By entering the origin and destination points, the app can calculate the

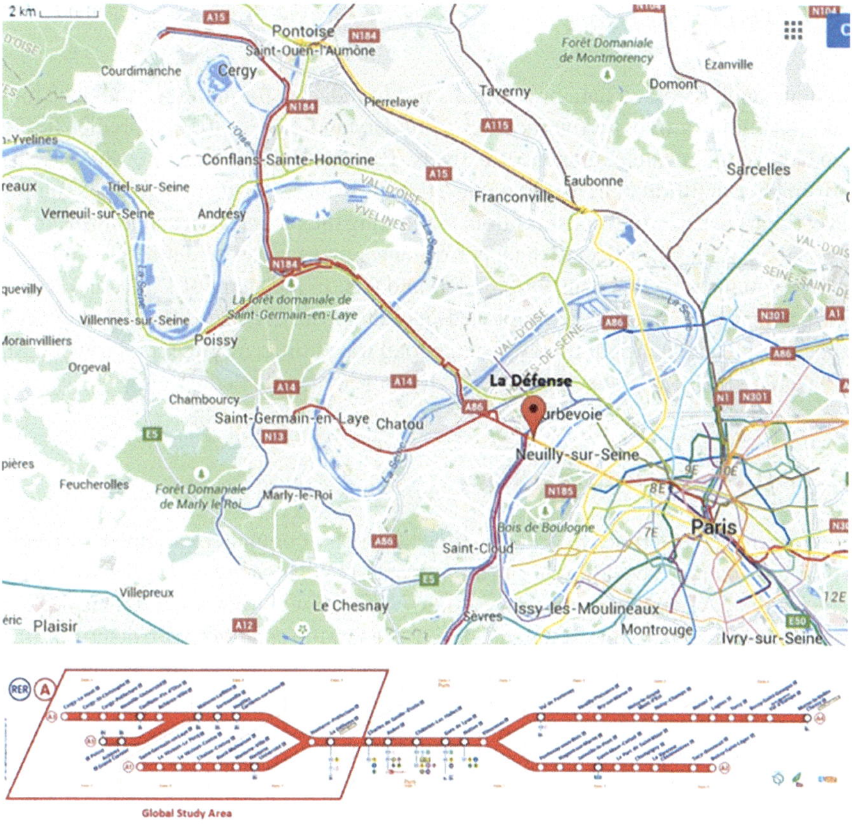

Fig. 3.3 Case study area in Paris, France (Liu et al. 2016)

optimal route, providing travel cost in euros, travel time, congestion time, and travel distance. PT data, including bus and railway timetables, are collected from openly published schedule by selecting options that minimized transfer waiting times.

To collect CB data, the shortest travel paths were identified. Initially, a grouping strategy was employed to cluster potential commuters into common trips via a 15-seat minibus. The primary criterion for grouping was the railway timetable, ensuring that people with similar origins and destinations shared the same CB trip. For profitability, minibuses required a minimum of six passengers. If this threshold was not met, CB service was not provided, necessitating re-grouping and forcing some commuters to take a train instead. Occasionally, individual trips could not be grouped due to significant discrepancies between expected and actual arrival times, requiring these travelers to use their PC or PT. After grouping, a nearest neighbor algorithm was applied to determine the minibus routes. Subsequently, Google Maps was used to estimate the total travel time for each route.

3.4.3 Data Processing

Accurate and processed data provides a solid foundation for planning and decision-making. It allows transport authorities to understand current trends, predict future demands, and allocate resources effectively. Processed data helps in evaluating the performance of different transport modes. It enables the assessment of metrics such as travel time, cost, and fuel consumption, facilitating the comparison and improvement of services. With the collected data, performance metrics can be defined to compare the performances of PC, conventional PT, and CB. In the study of Liu et al. (2016), four performance metrics are employed, including the average difference between expected and actual arrival times, average total travel time, average total travel cost, and average total fuel consumption, which are described in detail below.

3.4.3.1 Difference in Arrival Time

The expected arrival time (EAT) is the time at which a commuter anticipates arriving at their destination. The difference in arrival time (DAT) is calculated as the difference between the EAT and the actual arrival time (AAT). For conventional PT, the AAT is influenced by planned timetables and current road traffic conditions. To determine the DAT, the EAT is fixed for each passenger. Google Maps and Via Michelin Itinéraire (for Paris) are used to obtain the travel time and route for PC commuters. Similarly, by combining the timetables of buses and trains, the AAT for each commuter using conventional PT can be determined. The DAT is then calculated by comparing the EAT with the AAT for each mode of transport, as shown in Eq. (3.1).

$$DAT = |EAT - AAT| \tag{3.1}$$

3.4.3.2 Travel Time

The total travel time (TTT) may comprise several components, including in-vehicle travel time, out-of-vehicle waiting time, walking time, and transfer waiting time. For PC, the TTT is the sum of the driving time from home to the parking location (TT) and the walking time from the parking location to the workplace (WkT). Both times can be estimated using Google Maps. In the study of Liu et al. (2016), they conducted estimations from 7:00 A.M. to 10:00 A.M. based on each passenger's departure time, including an estimation of time lost in congestion. Similarly, they repeated the estimation for PT. Here, TT refers to the time spent on buses or trains. The WkT includes the walking time to bus/train stations, any potential walking time for transfers, and the walking time to the workplace. When transfers are involved, extra time may be required due to transfer waiting, denoted as WtT. They used

Google Maps to determine an itinerary, and transfer waiting time is calculated by simple subtraction.

For CB, travel data can be divided into two segments. Initially, commuters are picked up by minibuses and transported to the nearest train stations. The nearest neighbor algorithm is used to generate vehicle routes based on a predefined loading level. The TTT for CB is then estimated similarly to PT. For CB, TT includes time spent in the minibus and the train, while WkT corresponds to the travel time from the final train station to the workplace. Thus, the TTT for CB can be calculated using these components by:

$$TTT = TT + WkT + WtT \tag{3.2}$$

3.4.3.3 Travel Cost

For most commuters, travel cost (TC) is the primary factor influencing their choice of commuting mode. Travel costs can be calculated for the entire day, considering both morning and evening trips. Diesel is the predominant fuel used in New Zealand and France. Accordingly, the calculations are based on vehicles operating on diesel engines.

The Via Michelin app provided fuel consumption costs (FC) in euros for one-way trips. Additionally, the daily car parking price (CPP) is also included in the cost calculation. In Auckland, the vehicle distance traveled was estimated using Google Maps, allowing us to determine FC by multiplying the distance by the unit distance price. Thus, the calculation can be made by multiplying the distance and unit distance price to determine the FC and adding the estimated daily CPP.

$$TC = 2 \times FC + CPP \tag{3.3}$$

The following distance tariff scheme that includes two fare components: variable fare and constant fare, is used to calculate the CB fare cost. The variable fare can be calculated by:

$$F_1 = max\left\{0, F\prime \cdot \lceil \frac{L - L_0}{L_1} \rceil \right\} + F_0 \tag{3.4}$$

where L is the length of a trip, F_0 is the basic fare that is compulsorily charged as long as one uses the CB service, $F\prime$ is the fare factor employed for calculating fares for different trip lengths, L_0 and L_1 are the threshold length and length factor respectively, and function $\lceil x \rceil$ is the celling function, which gives the smallest integer $\geq x$. According to this definition, when the length of a trip is less than the threshold length, L_0, the charge will only be for the basic fare F_0.

Combining the variable fare with the constant fare, the total fare can be calculated by:

$$F = \alpha_1 \cdot F_1 + \alpha_2 \cdot F_2 \tag{3.5}$$

where F_2 is the constant fare, α_1, $(0 \leq \alpha_1 \leq 1)$ and α_2, $(0 \leq \alpha_2 \leq 1)$ are the discount factors of the constant fare and the variable fare, respectively.

3.4.3.4 Fuel Consumption

Fuel consumption (FC) is calculated per liter per person. For cars in Paris, the Via Michelin app can be used to obtain the FC in euros per person. For Auckland, in the study of Liu et al. (2016), they used an average fuel consumption of 8.61 L per 100 km and multiplied it by the distance traveled. For PT, they estimated fuel consumption at a rate of 0.45 L per kilometer. This value was then multiplied by the distance traveled (D) and divided by the average number of passengers on a bus, which estimated to be 10 people. The FC for PT is calculated using the following formula:

$$FC_{pt} = \frac{0.45 \times D}{10} \tag{3.6}$$

This approach allows for accurate estimation of fuel consumption for both PCs and PT in various urban settings. In the study by Liu et al. (2016), the fuel consumption of customized buses (CB) is estimated at 0.26 L per kilometre, reflecting the average consumption of a minibus. The calculation method parallels that used for PT but considers minibuses with 15 seats, assuming an average of seven passengers on board. Thus, the fuel consumption (FC) for PT is calculated as follows:

$$FC_{cb} = \frac{0.26 \times D}{7} \tag{3.7}$$

These methodologies ensure precise fuel consumption estimates for different commuting modes, aiding in the evaluation and comparison of their efficiency and sustainability in urban transport systems.

3.5 Results and Discussions

Liu et al. (2016) applied the methodological framework and performance metrics to a case study of the city of Auckland and Paris. First, 100 candidate commuting trips were randomly generated for the two case study areas. Data on travel time, travel cost, and fuel consumption were collected for all these trips. Average values were then calculated to facilitate comparisons among the three different commuting modes: private car, traditional public transport, and customized bus. Some interesting results are obtained.

For the case study in Auckland, the 100 candidate commuting trips are classified into eight groups. The average results for these eight groups are graphically depicted in Fig. 3.4. Generally, in Auckland, PT and CB exhibit similar performance compared to PC. Notably, CB is less expensive than both PT and PC, likely due to the high cost of PT services in Auckland. PCs have the shortest average travel time and the lowest average difference in arrival times. However, they consume significantly more fuel and incur higher costs compared to PT and CB. While CB consumes slightly more fuel than traditional PT, it offers considerably greater comfort by eliminating congestion and transfer waiting, and ensuring available seats. It is essential to note that this initial study is limited by the number of candidate trips analyzed. With a larger sample of commuter trips, the efficiency of CB compared to PC and PT would likely be more pronounced.

For the case study in Paris, the 100 candidate commuting trips are classified into twelve groups. The average results of these twelve groups are depicted in Fig. 3.5. It shows that PCs have the lowest deviation from expected arrival time, while CBs have the highest deviation. This could be attributed to the limited and dispersed sample of trips, which increases the time required to pick up commuters from their residences. However, despite CBs having a larger deviation in arrival times compared to PCs, they are significantly cheaper and consume much less fuel. Furthermore, when comparing CBs to PT, CBs are 15% faster but also 33% more expensive. This cost difference may be due to reduced transfer times with CBs. To further reduce the average travel cost and fuel consumption for CB, expanding the sample size of commuting trips would be beneficial.

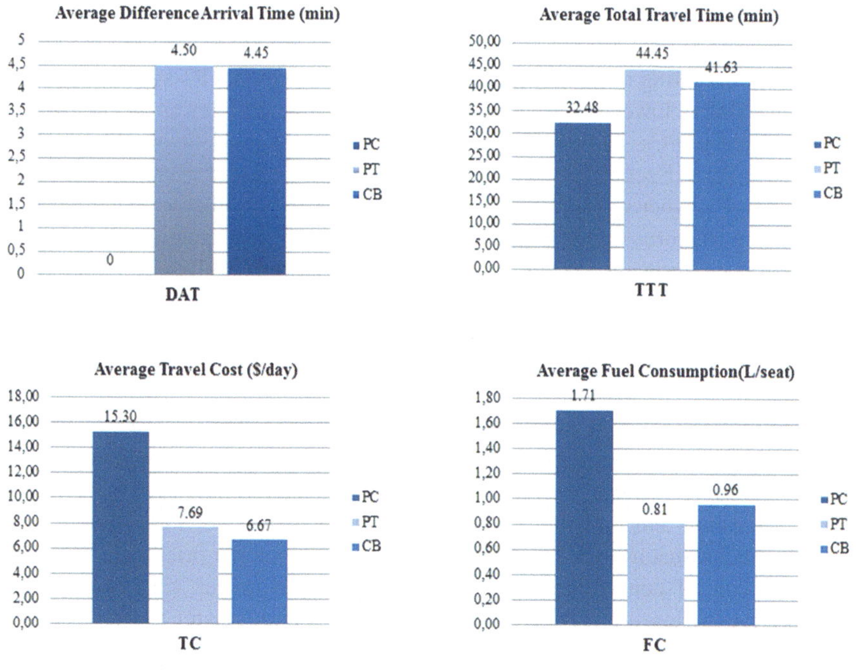

Fig. 3.4 Average performance metrics measured per mode in Auckland (Liu et al. 2016)

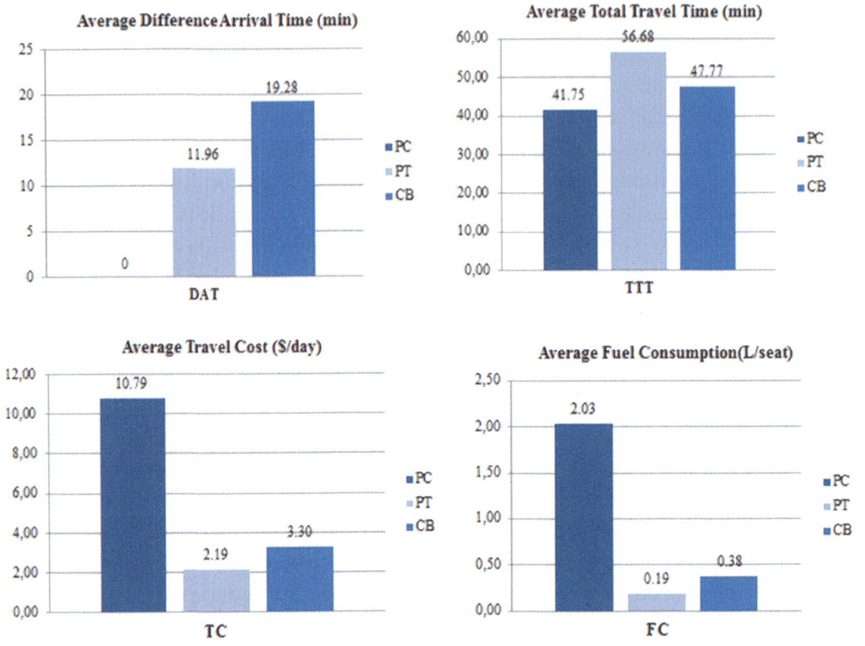

Fig. 3.5 Average performance metrics measured per mode in Paris (Liu et al. 2016)

Based on the case study results from Auckland and Paris, CB emerges as a promising commuting mode compared to PCs and conventional PT systems. In both cities, CB exhibits lower average travel costs and significantly lower fuel consumption compared to PCs. However, its performance appears more favorable in Auckland than in Paris. Specifically, in Auckland, CB shows a lower average travel cost compared to PT, whereas in Paris, PT tends to be more cost-effective than CB. This difference underscores the varying suitability of CB across different urban contexts.

Moreover, in Paris, the average DAT for CB is twice that of PT, which may raise concerns among commuters. This discrepancy is partly due to how CB groups are formed and the limited sample size in our study. Additionally, the superior efficiency of the PT network in Paris compared to Auckland contributes to the stronger performance of PT in Paris.

Currently, there is a lack of investment in enhancing the efficiency of Auckland's PT systems, which contributes to the prevalent use of PCs. This study underscores that CB can serve as a viable alternative to both PCs and PT, potentially alleviating traffic congestion and reducing commuting costs in Auckland. Furthermore, our findings indicate that using PT or CB in Auckland offers similar pricing, travel time, and DAT, with PT consuming 15% less fuel than CB. However, CB enhances the commuting experience by reducing walking, waiting, and transfer times. If efforts are made to reduce minibus fuel consumption and maintain competitive pricing relative to PT, CB could emerge as an attractive solution in Auckland. In Paris, while

CB's fuel consumption and travel costs are slightly higher than PT's, its potential to outperform PCs and PT remains contingent upon broader participation and operational enhancements. The case study results show CB has the potential to alleviate congestion and reduce commuting costs, while enhancing the commuting experience through reduced wait and transfer times.

3.6 Conclusions

With urban traffic congestion exacerbated by the widespread use of private vehicles, governments are increasingly seeking viable alternatives. PT systems often fall short in terms of user satisfaction and service flexibility, prompting interest in CB as a novel commuting mode. This chapter aims to compare CB's performance with PT and PC to ascertain its potential benefits. A methodological framework was described to quantify operational metrics such as travel cost, travel time, and fuel consumption, which were evaluated across different cities.

Case study results from Auckland and Paris reveal that PCs offer the fastest and most reliable commute, albeit at the double the cost and significantly higher fuel consumption compared to PT and CB in both cities. CB shows potential profitability in Auckland relative to Paris, where PT's superior infrastructure results in lower travel costs and arrival time discrepancies compared to CB. In Auckland, PT and CB exhibit similar average travel times and arrival discrepancies, making CB financially attractive due to lower fares.

Despite CB's higher fuel consumption compared to PT, its enhanced comfort features, including reduced walking and transfer times, and less crowded vehicles, contribute to a more comfortable commuting experience. Moreover, an increasing number of CB agencies are adopting electric vehicles to reduce fuel costs and overall emissions. This chapter concludes that CB represents a viable commuting alternative across different city contexts, with opportunities for further efficiency improvements through future research, particularly focusing on the impact of electric minibuses on CB system performance.

References

AASHTO. 2015. Commuting in America 2013: The national report on commuting patterns and trends [WWW Document]. http://traveltrends.transportation.org/Pages/default.aspx.

Anable, J. 2005. 'Complacent car addicts' or 'aspiring environmentalists'? Identifying travel behaviour segments using attitude theory. *Transport Policy (Oxford)* 12: 65–78. https://doi.org/10.1016/J.TRANPOL.2004.11.004.

Beirão, G., and J.A. Sarsfield Cabral. 2007. Understanding attitudes towards public transport and private car: A qualitative study. *Transport Policy (Oxford)* 14: 478–489. https://doi.org/10.1016/j.tranpol.2007.04.009.

Ceder, A. 2016. *Public transit planning and operation: Modeling, practice and behaviour, Second Edition*. Public Transit Planning and Operation. https://doi.org/10.1201/B18689.

Ceder, A. (Avi). 2021. Urban mobility and public transport: Future perspectives and review. *International Journal of Urban Sciences* 25: 455–479. https://doi.org/10.1080/12265934.2020.1799846

Chen, E., Z. Ye, and H. Wu. 2021. Nonlinear effects of built environment on intermodal transit trips considering spatial heterogeneity. *Transportation Research Part D: Transport and Environment* 90: 102677. https://doi.org/10.1016/J.TRD.2020.102677.

Galdames, C., A. Tudela, and J.A. Carrasco. 2001. Exploring the role of psychological factors in mode choice models by a latent variables approach.

Hiscock, R., S. Macintyre, A. Kearns, and A. Ellaway. 2002. Means of transport and ontological security: Do cars provide psycho-social benefits to their users? *Transportation Research Part D: Transport and Environment* 7: 119–135. https://doi.org/10.1016/S1361-9209(01)00015-3.

Kingham, S., J. Dickinson, and S. Copsey. 2001. Travelling to work: Will people move out of their cars. *Transport Policy (Oxford)* 8: 151–160. https://doi.org/10.1016/S0967-070X(01)00005-1.

Kuhnimhof, T., B. Chlond, S. Von Der Ruhren. 2006. Users of transport modes and multimodal travel behavior, 40–48. https://doi.org/10.1177/0361198106198500105

Liu, J., B. Wang, and L. Xiao. 2021. Non-linear associations between built environment and active travel for working and shopping: An extreme gradient boosting approach. *Journal of Transport Geography* 92: 103034. https://doi.org/10.1016/J.JTRANGEO.2021.103034.

Liu, T., and A. Ceder. 2015. Analysis of a new public-transport-service concept: Customized bus in China. *Transport Policy (Oxford)* 39: 63–76. https://doi.org/10.1016/J.TRANPOL.2015.02.004.

Liu, T., A. Ceder, R. Bologna, and B. Cabantous. 2016. Commuting by customized bus: A comparative analysis with private car and conventional public transport in two cities. *Journal of Public Transportation* 19: 55–74. https://doi.org/10.5038/2375-0901.19.2.4.

Sandow, E. 2011. On the road: Social consequences of commuting long distances to work.

Stutzer, A., and B.S. Frey. 2008. Stress that doesn't pay: The commuting paradox. *Scandinavian Journal of Economics* 110: 339–366. https://doi.org/10.1111/J.1467-9442.2008.00542.X.

Tertoolen, G., D. Van Kreveld, and B. Verstraten. 1998. Psychological resistance against attempts to reduce private car use. *Transportation Research Part A: Policy and Practice* 32: 171–181. https://doi.org/10.1016/S0965-8564(97)00006-2.

Tong, L. (Carol), L. Zhou, J. Liu, and X. Zhou (2017). Customized bus service design for jointly optimizing passenger-to-vehicle assignment and vehicle routing. *Transportation Research Part C: Emerging Technologies* 85: 451–475. https://doi.org/10.1016/J.TRC.2017.09.022

Part II
Travel Behavior of Customized Bus Users

Chapter 4
Customized Bus Choice Behavior and User Preference Heterogeneity

Abstract Demand-response strategies for customized buses (CB) typically treat different passengers homogeneously, ignoring the heterogeneity between passengers in their choice preferences. In order to gain further insights into people's choice behavior in CB and explore the influence of latent variables, this study combined the characteristics of CB operation, and latent variables reflecting convenience, comfort and reliability into a Mixed logit model to form a hybrid choice model. A survey was conducted on 30 CB routes in Dalian to study the operational characteristics of CB and quantitatively analyze the impact of various attributes on CB choice behavior using empirical data. The results show that CB is more attractive to middle-aged and young women with middle to high incomes. Reliability and traditional variables such as travel time have an equally important impact on choice, and the utility of average cognitive reliability is equivalent to the utility of reducing travel time by 6.15 min. The probability of choosing CB will increase by 1.09 times for every increase in the perceived reliability evaluation level. The hybrid choice model has a higher goodness of fit than the traditional logit model, and the prediction results are more robust. Travelers have heterogeneous preferences for travel time, waiting time, seat availability on regular buses, discounts on fares, and detour rates. To ensure a stable full-load rate, a discount of at least 12% on fares is needed to offset the perceived negative utility of approximately 90% of travelers, accounting for heterogeneity. Therefore, on the basis of providing multi-vehicle types and multimodal travel services, operators need to find a balance point between demand-responsive customized bus stops and route adjustments, and fare adjustments.

Keywords Traffic engineering · Travel mode choice · Hybrid choice model · Customized bus · Latent variable

J. Wang et al., *Customized Bus: User Travel Behavior and Demand Evolution*, Urban Sustainability, https://doi.org/10.1007/978-981-96-3488-0_4

4.1 Introduction

Diversified public transportation is considered an important approach to alleviate traffic congestion and enhance citizens' travel accessibility. With the rapid development of transportation systems, the quality of multimodal public transportation services has been continuously improving (Yun et al. 2017). The disconnection between transportation supply and travel demand is a major issue faced by traditional public transportation. The rapid development of information technology has provided new opportunities for balancing supply and demand. A novel public transportation service model, known as customized bus (CB) service, has been flourishing in China. Compared to traditional buses, CB service provides high-quality and personalized travel services for travelers with specific needs, such as commuters. Features such as individual seating, direct travel without transfers, and responsive service during peak hours offer travelers a superior travel experience (Wang et al. 2018). The successful implementation of CB service has brought positive impacts on optimizing urban transportation network structures, alleviating traffic pressure, and reducing traffic pollution. CB service has been operating in China for nearly a decade, yet it still faces challenges such as insufficient attractiveness, low occupancy rates, high operational costs (Fu et al. 2019), and low passenger loyalty (Wang et al. 2020a). The demand-responsive operational strategies and dynamically optimized design of CB service have also been subject to considerable scrutiny. The willingness of residents to continuously use CB service over the long term requires further exploration.

To explore the influencing mechanism of passengers' choice of CB as their travel mode, Jian et al. (2019) conducted an in-depth analysis of CB's operational status, demand characteristics, and service features. By employing a structural equation model, they initially identified the influencing mechanism of residents' willingness to choose CB for their travel needs. Using both Stated Preference (SP) and Revealed Preference (RP) surveys conducted in Harbin, Cao and Wang (2016b) determined key factors that influence the choice of CB through logistic regression analysis, including car ownership, commuting distance, and travel satisfaction. Zhang and Jiao (2020) focused on enhancing the travel experience of bus users, targeted white-collar individuals with above-average income and designed a pricing mechanism tailed to the actual situation. Based on actual operational data of CB in Dalian, Wang et al. (2019b) demonstrated significant improvements in service levels compared to regular bus services and showcased CBs' ability to complement the public transportation system in areas where regular bus service is inadequate. They further investigated key factors influencing user loyalty, helping CB operators identify and quantify operational risk factors, and provided recommendations such as reducing detours and improving operational speed (Wang et al. 2020a). Additionally, data-driven research has provided a more comprehensive understanding of demand evolution. Chen et al. (2020) established a framework for analyzing the evolution of CB demand using multiple data sources. They identified latent demand from bus and internet users, proposed methods for integrating travel demands and selecting station locations, and achieved city-scale demand mining and fully responsive CB route design.

To investigate the mechanism influencing passengers' continued choice of CB, this study aims to explore individual preference differences and provide more accurate demand estimation for dynamic response operational strategies and route adjustment (Li et al. 2021). Previous studies have shed light on significant individual effects in the choice of CB users (Schlüter et al. 2021), as well as the evolution of mode share based on the heterogeneous patterns of traveler behavior over time (Anable 2005). However, existing research mainly focuses on the influence of travel mode attributes and individual characteristics on choice outcomes, offering a theoretical basis for the early-stage operation of CB service. In practice, driven by demand-responsive operational models and profit-driven strategies, adjustments in routes, stations, and operational timing are necessary to meet passengers' diverse needs. Failure to meet these needs may result in passengers switching to other modes or even the cancellation of certain CB routes.

Individual preferences and their differences reflect different values, social norms, and lifestyles. These preferences are often not directly observable, but understanding the decision-making process can be improved by investigating individuals' underlying attitudes (Hess et al. 2018). Neglecting latent variables can lead to biased parameter estimates (Chen and Gan 2014). Hybrid Choice Models (HCM) combine factor analysis from discrete choice models with attitude latent variable quantification from structural equation models, and they have been widely used in travel choice research (Ju et al. 2017). Domestic and international studies have extensively explored the influence of these psychological latent variables on travel decisions using the HCM framework. They have investigated daily travel choices (Jing et al. 2014), commuting mode choices (Cheng et al. 2016), and school travel mode choices considering the overall psychological factors of family members (Jing et al. 2017). Based on the Technology Acceptance Model (TCM) framework, Qi et al. (2021) extracted four categories of latent variables: ride-hailing habits, perceived reliability of ride-hailing platforms, social influence of ride-hailing, and traveler's preference for altruism. They constructed a structural equation model to examine the intention to use autonomous ride-hailing cars. Paulssen et al. (2014) applied the hybrid choice model to describe the influence of values and attitudes on observable individual travel behavior, confirming the hierarchical causal representation of values-attitudes-behavior. Raveau et al. (2012) validated improvement strategies and effects of parameter estimation in the hybrid choice model from a practical application perspective, thereby enhancing the model's applicability. These studies highlight the importance of selecting appropriate latent variables when constructing hybrid choice models.

By incorporating psychological latent variables into the traditional discrete choice modeling framework, richer information can be provided for the study of decision mechanisms. In this study, we first construct psychological latent variables that influence the choice of CB service. We use the Multiple Indicators Multiple Causes (MIMIC) model to depict the relationship between individual economic attributes, measurement indicators, and unobservable latent variables, and integrate these psychological latent variables into the travel choice model. Furthermore, we overcome the limitations of the traditional multinomial logit model, which assumes independence and unrelatedness of attributes, by using the more flexible Mixed logit model to quantify heterogeneity in travelers' preferences. We assess the sensitivity of different passengers to dynamic scenarios and predict the impacts of factors such as route adjustments, station settings, and fare fluctuations.

4.2 Methodology

4.2.1 Hybrid Choice Model (HCM)

This study adopts an integrated Hybrid Choice Model (HCM) proposed by Ben-Akiva et al. (2002) for analyzing the choice of CB service. The theoretical framework is depicted in Fig. 4.1. In the rational utility of the traditional discrete choice model, latent variables (Jing et al. 2014) that influence the decision-making process are incorporated to synthesize the perceived utility. The relationships between individuals' socio-economic attributes, travel mode attributes, and latent variables with the travel choice are identified by the perceived utility function, aiming to capture the impacts of explanatory variables on choice preferences. The ellipses in the figure represent unobservable psychological factors, i.e., latent variables. The rectangles represent directly observable exogenous variables, including individuals' socio-economic attributes, travel mode attributes, and measurement indicators. The solid arrows indicate structural relationships that reflect causal connections, while the dashed arrows represent measurement relationships.

The explanatory variables in the perceived utility function of travelers can be represented as follows:

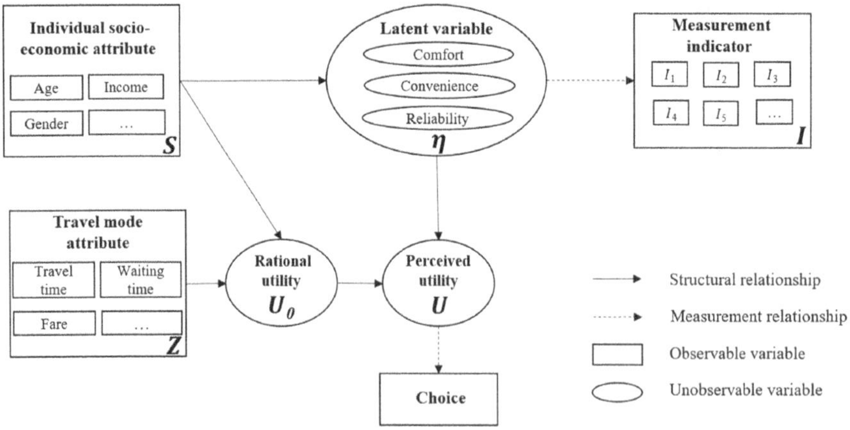

Fig. 4.1 Hybrid choice model

$$X_{ni} = \begin{bmatrix} S_n \\ Z_i \\ \eta_n \end{bmatrix} \tag{4.1}$$

where X_{ni} is the known complete set of explanatory variables, S_n is the vector of observable socio-economic attributes of the traveler, Z_i is the vector of observable attributes of the travel mode, η_n and is the vector of latent variables of the traveler, which includes three dimensions: comfort, convenience, and reliability perceived by CB passengers.

4.2.2 Multiple Indicators Multiple Causes (MIMIC) Model

The MIMIC model is used to depict the relationship between individual socio-economic attributes, measurement indicators, and unobservable latent variables. The model consists of two parts: the measurement model and the structural model. The measurement model establishes the relationship between socio-economic attributes and latent variables, while the structural model uses observable characteristics of individuals to explain the latent variables (Chen and Gan 2014). The model can be formulated as follows:

$$\eta = \Gamma S + \xi \tag{4.2}$$

$$I = \Lambda \eta + \upsilon \tag{4.3}$$

Equation (4.2) represents the structural model, where η is the vector of latent variables, S is the vector of exogenous observable variables that influence the latent variables, and $\boldsymbol{\Gamma}$ is the parameter matrix to be estimated. represents measurement errors. Equation (4.3) represents the measurement model, where I is the vector of observable indicators of the latent variables, $\boldsymbol{\Lambda}$ is the parameter matrix to be estimated, and υ is the error term.

4.2.3 Mixed Logit Model

The most widely used model in discrete choice modeling is the logit model, which assumes the independence of irrelevant alternatives (IIA) property (Hess and Train 2017). To avoid the bias caused by the IIA property, this study adopts the Mixed logit model in the discrete choice modeling part. The Mixed logit model assumes that the parameters to be estimated follow a probability distribution, thereby explaining the heterogeneity. In this study, each traveler is faced with choosing the optimal travel decision among 6 stated preference (SP) scenarios. Using the Mixed logit model can better address the issue of panel correlation between multiple choices made by respondents. When an individual faces the choice, the general form of the utility expression is as follows:

$$U_{ni} = V_{ni} + \varepsilon_{ni} = \sum_{k=1}^{K} \beta_{nik} x_{nik} + \varepsilon_{ni} = \beta_n X_{ni} + \varepsilon_{ni} \quad (4.4)$$

where, U_{ni}、V_{ni}、ε_{ni} represent the utility, fixed utility, and random utility of the individual n choosing travel mode i, respectively. x_{nik} represents the explanatory variable for the choice of travel mode, which includes the individual socio-economic attributes and specific attributes of the travel mode. β_{nik} represents the corresponding parameter for the explanatory variable, and K represents the number of explanatory variables. X_{ni} represents the vector of complete explanatory variables that can be observed, and β_n represents the corresponding parameter vector.

The Mixed logit probability is the integral of the standard logit probability over the parameter probability density function. The probability of individual choosing candidate option is:

$$P_{ni} = \int \frac{\exp(\beta_n X_{ni})}{\sum_{i=1}^{I} \exp(\beta_n X_{ni})} f(\beta_n | X_{ni}) d\beta_n \quad (4.5)$$

where $f(\beta_n | X_{ni})$ represents the probability density function of the parameter for the explanatory variables β_n, given the complete set of observed explanatory variables X_{ni}. I represents the number of candidate travel modes.

This study employs the method of continuous numerical estimation, which consists of two steps: (1) estimating the parameters of the MIMIC model using maximum likelihood estimation, and (2) using the estimated parameters from step 1 to calculate the fitted values of latent variables. These fitted values are then used as exogenous variables in the discrete choice model, and the model parameters are estimated using maximum likelihood estimation as well.

4.3 Data Collection and Latent Variable Estimation

4.3.1 Survey

The first CB route in Dalian started operating in March 2014. The operator implemented flexible operational strategies, adjusting, merging, or canceling routes with consecutive occupancy rates below 50%. The survey for this study was conducted from July 3rd to July 5th, 2017. As of the survey date, there were a total of 148 CB routes covering 1556 stations in the city. For the purpose of this study, 30 randomly selected operational routes were surveyed.

The survey questionnaire consisted of three parts. The first part collected socio-economic attributes of the travelers, including gender, age, number of vehicles owned, monthly income, and monthly commuting expenses. The second part consisted of Stated Preference (SP) surveys for travel choices. Based on the actual operational conditions of CB in Dalian, three travel distances were set, and 18 CB and regular bus travel scenarios were generated using an orthogonal experimental design, as shown in Table 4.1. To avoid multicollinearity among attributes, three sets of questionnaires were constructed, each containing six travel choice scenarios.

The third part focused on measuring psychological factors. Taking into account the research on psychological variables influencing travel choices by Vredin Johansson et al. (2006) and the travel characteristics of CB, three latent variables were considered: comfort, reliability, and convenience. Each latent variable corresponded to 4 to 5 questions, and a Likert scale with five points was used, with scores ranging from 1 to 5 representing "strongly disagree," "disagree," "neutral," "agree," and "strongly agree," respectively.

A total of 586 surveys were collected, of which 446 were valid, resulting in a valid response rate of 76%. Based on the Events Per Variable (EPV) method (Peduzzi et al. 1996), the sample size was determined to meet the modeling requirements.

The descriptive statistical analysis of the surveyed individuals' socio-demographic characteristics is presented in Table 4.2. Currently, the majority of Dalian's CB passengers are middle-income young women. Among the surveyed individuals,

Table 4.1 The attributes of travel survey

Variable type	Variable	Mode	Number of levels	Level values
Attributes shared by travel modes	Waiting time/min	CB	3	3、8、15
		PT	3	5、13、20
	Travel time/min	CB	10	30、40、45、50、60、65、70、75、80、90
		PT	10	40、45、50、60、65、70、75、80、90、100
	Fare /yuan	CB	16	4.5、6、6.4、7、8、8.8、9、10.8、12、12.8、13、13.5、14.4、15、15.3、18
		PT	4	2、3、4、5
	Distance/km	CB	3	10、20、25
		PT	3	10、20、25
Attributes specific to PT	Seat availability	PT	2	Yes、No
	Number of transfers	PT	3	0、1、2
Attributes specific to CB	Detour distance/km	CB	3	0、3、5
	Discount rate	CB	3	100%、90%、80%

83.18% are female, as both vehicle ownership and driver's licenses among women are lower than those among men (Wang et al. 2019b). More than 90% of the surveyed individuals are between the ages of 19 and 40. Although all surveyed individuals are passengers of CBs, the results show that the vast majority of passengers' families own one or more private cars. Nearly half of the surveyed individuals have monthly incomes ranging from 5001 to 10,000 yuan, with an average monthly income of 6807 yuan, slightly higher than the reported average monthly income of 6446 yuan by the Dalian municipal government in 2017.

Latent variables are typically measured by multiple indicators, and the statistical results are presented in Table 4.3. Passengers are generally satisfied with the comfort, convenience, and reliability of CBs, with only four measurement indicators having means below 4.

In terms of comfort measurement indicators, the highest satisfaction is with the level of crowding inside the vehicle, followed by "quiet and noise-free interior." This highlights a significant difference between CB and regular buses, indicating that the "one person per seat" travel mode significantly enhances comfort, which is a key factor in attracting passengers. However, satisfaction with the smoothness of vehicle operation is relatively low, suggesting that improving comfort should focus on the stability of the vehicle during travel.

Table 4.2 Descriptive statistics

Symbol	Individual attribute	Indicator	Percentage
S_1	Gender	Male	16.82%
		Female	83.18%
S_2	Age	19–30	39.01%
		31–40	56.28%
		41–50	4.04%
		More than 51	0.67%
S_3	Number of private cars owned	0	30.04%
		1	62.78%
		More than 2	7.17%
S_4	Income	Less than 3000 yuan	2.24%
		3001–5000 yuan	38.34%
		5001–10,000 yuan	46.86%
		10,001–15,000 yuan	10.31%
		15,001–20,000 yuan	1.79%
		More than 20000yuan	0.45%
S_5	Commuting Expenses	0–100 yuan	1.35%
		100–200 yuan	6.28%
		200–400 yuan	67.71%
		More than 400 yuan	24.66%

Table 4.3 Measurement indicators

Latent variable	Variable	Problem description	Mean	Standard Deviation
Comfort	I_1	Not crowded	4.25	0.85
	I_2	No noise	4.11	0.90
	I_3	Clean and tidy	4.08	0.89
	I_4	Comfortable air conditioning temperature	3.98	1.02
	I_5	Smooth driving	3.70	1.20
Convenience	I_6	Convenient arrival at stops	3.96	1.27
	I_7	Able to use bus lanes	4.49	0.56
	I_8	Convenient use of ticket purchase apps	4.07	1.08
	I_9	Timely response	3.43	1.42
Reliability	I_{10}	Accurate arrival times	4.14	0.83
	I_{11}	Few vehicle breakdowns	4.13	0.79
	I_{12}	Fewer instances of being late for work	4.29	0.67
	I_{13}	Timely notifications of vehicle arrivals	4.43	0.66

Among the convenience measurement indicators, passengers are most satisfied with the fact that "CB can use bus lanes," which also receives the most consistent ratings. This underscores the advantage of CB during peak hours compared to private cars and taxis. However, passenger response time and the distance to walk to bus stops are both below the satisfaction level, and the ratings are dispersed. This suggests that the flexibility and coverage of CB need further improvement. More reasonable site selection and route planning remain key factors in enhancing the attractiveness of customized buses in Dalian. Additionally, this reflects the differences in passenger preferences for CB service quality.

Reliability measurement indicators have received consistently high ratings with relatively narrow rating ranges. This indicates that CB in Dalian have a significant advantage in terms of on-time performance and a low accident rate during peak hours.

4.3.2 *Latent Variable Estimation*

To ensure the reliability and stability of data collection, a reliability and validity analysis was conducted on the collected questionnaires (Hörcher et al. 2017). The results of Cronbach's Alpha, the KMO sample measure, and Bartlett's sphericity test are detailed in Appendix A. The data passed the reliability and validity tests, meeting the modeling requirements. Using AMOS, a path diagram (Fig. 4.2) was established to illustrate the relationships between the observed variables and latent variables. The

MIMIC model, which includes latent variables, was estimated using the maximum likelihood estimation method, and the final estimated results were obtained. Specific model fit indices can be found in Appendix B, and all indices met the adequacy standards, indicating a good fit. Relationships between variables with a confidence level of 90% were retained, and the standardized path coefficient fit results for each variable are presented in Table 4.4.

In the MIMIC model, the values of latent variables are directly proportional to the measurement indicators, meaning that higher passenger satisfaction corresponds to larger values of these latent variables. An analysis of the structural relationship between latent variables and the socioeconomic attributes of travelers reveals several key findings. As age increases, passengers tend to rate CB services higher in terms of comfort, convenience, and reliability. This suggests that older passengers have a higher level of satisfaction with these aspects of CB travel. Travelers who own private cars tend to give lower ratings to all three latent variables related to CB services. This implies that car owners may have different expectations or preferences when it comes to CB services, resulting in lower satisfaction levels. Passengers with higher monthly incomes tend to rate convenience and reliability lower. This suggests that individuals with higher incomes may have more demanding expectations regarding the convenience and reliability of CB services. Passengers with higher monthly commuting expenses tend to rate comfort and convenience lower. This indicates that individuals who spend more on their daily commute have higher expectations regarding the comfort and convenience of CB services. They may expect a higher level of service quality to justify their commuting expenses.

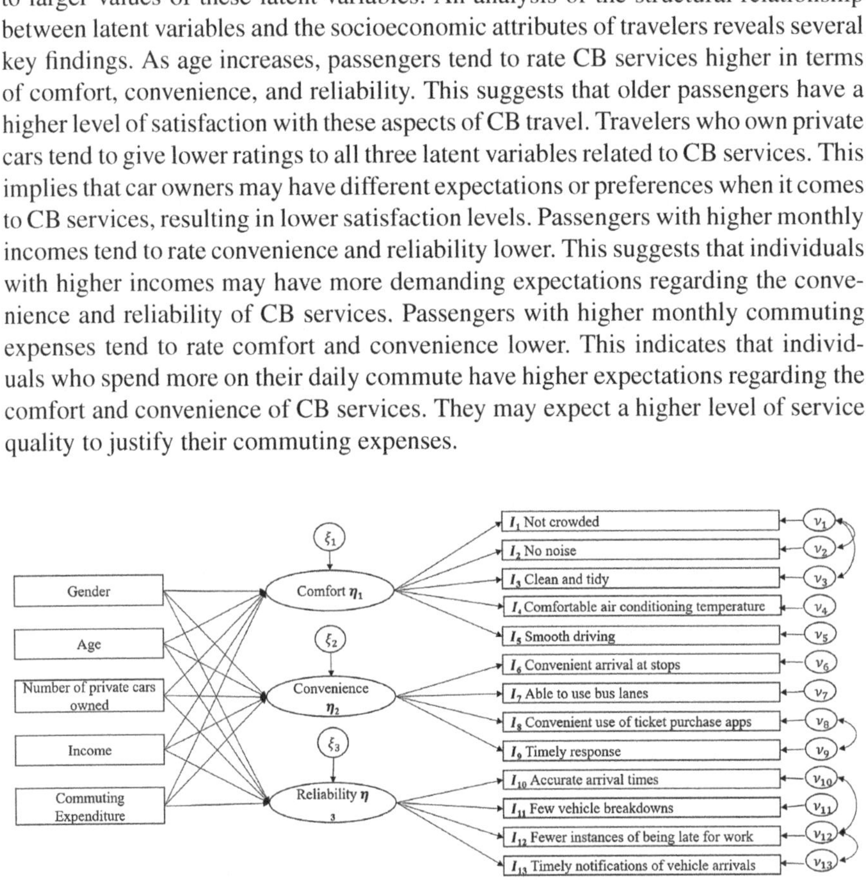

Fig. 4.2 Path diagram of the relationship between explicit and latent variables

Table 4.4 Estimation results of MIMIC model

	Standardized path coefficient	Standard error	t-value
Comfort <--- Gender	0.045	0.035	1.280
Comfort <---Age	0.005**	0.002	2.080
Comfort <--- Number of private cars owned	-0.145***	0.029	-5.100
Comfort <---Income	-0.006	0.004	-1.369
Comfort <--- Commuting Expenditure	-0.061***	0.017	-3.623
Convenience <--- Gender	-0.008	0.033	-0.229
Convenience <--- Age	0.009***	0.002	4.029
Convenience <--- Number of private cars owned	-0.111***	0.027	-4.111
Convenience <--- Income	-0.008*	0.004	-1.811
Convenience <--- Commuting Expenditure	-0.081***	0.016	-5.024
Reliability <--- Gender	0.123***	0.042	2.895
Reliability <--- Age	0.007**	0.003	2.436
Reliability <--- Number of private cars owned	-0.134***	0.035	-3.886
Reliability <--- Income	-0.019***	0.005	-3.63
Reliability <--- Commuting Expenditure	-0.005	0.021	-0.257

Note *** represents $p < 0.01$; ** represents $p < 0.05$; * represents $p < 0.1$

4.4 Analysis of HCM Results

Based on the estimation results of the MIMIC model described above, predictions were made for the latent variables of CB passengers. Furthermore, these latent variables were used as exogenous variables in the mixed-choice model. The analytical results were compared between the binary logit model without latent variables and the HCM model with latent variables. This analysis aimed to uncover the specific impact of individual preference heterogeneity on demand-responsive operational methods.

4.4.1 HCM Calibration Results

We established both a binary logit model without latent variables and a mixed-choice model that includes latent variables using the R programming package "mlogit", which has been proven to efficiently analyze various discrete choice models, including the Mixed Logit model. Taking regular public transit as the reference option, we considered travelers' personal attributes, travel mode attributes, and latent variables. The utility variables were defined as shown in Table 4.5. Based on Eqs. (4.3) and (4.5), we obtained the total utility for traveler n choosing regular public transit (PT) and CB, respectively:

$$U_{nPT} = \beta_{nWT} WT_{PT} + \beta_{nTT} TT_{PT} + \beta_{nTC} TC_{PT} + \beta_{nST} ST_{PT} + \beta_{nTF} TF_{PT} + \varepsilon_{nPT} \tag{4.6}$$

$$\begin{aligned} U_{nCB} = \beta_0 + \beta_{nWT} WT_{CB} + \beta_{nTT} TT_{CB} + \beta_{nTC} TC_{CB} + \beta_{nDP} DP_{CB} + \beta_{nDC} DC_{CB} \\ + \beta_{GD} GD + \beta_{AG_1} AG_1 + \beta_{AG_2} AG_2 + \beta_{CR} CR + \beta_{MI_1} MI_1 + \beta_{MI_2} MI_2 \\ + \beta_{DP} DP + \beta_{\eta_1} \eta_1 + \beta_{\eta_2} \eta_2 + \beta_{\eta_3} \eta_3 + \varepsilon_{nCB} \end{aligned} \tag{4.7}$$

The Mixed Logit model was applied using the Halton sequence method, with 250 random samplings for simulation probability calculations. The model estimation results, with regular public transit as the base case, are shown in Table 4.6. The McFadden R^2 value ranges between 0 and 1, with a value around 0.2 generally considered indicative of high model accuracy. For the Binary Logit model, the McFadden R^2 value was only 0.043, while the McFadden R^2 value increased to 0.229 for the Mixed Choice model. This substantial improvement in fit indicates that individual heterogeneity and latent psychological variables have a significant impact on travelers' choices and cannot be ignored.

4.4.2 Results Comparison Analysis

(1) The Impact of Individual Socio-economic on Travel Mode Choice.

Both the BLM and HCM yield consistent estimation results. Compared to males, females show a greater inclination towards choosing CB for travel. The coefficient estimated for the middle-aged group, with reference to those aged 40 and above, is negative, indicating higher instability in their continued selection of CB. The group with private cars has a negative impact on the total utility, suggesting that travelers without private cars are more willing to choose CB. With the reference group being individuals with a monthly income exceeding 10,000 yuan, those with lower and moderate incomes have a lower probability of choosing CB, and the probability decreases with lower income levels. Additionally, travelers with a higher proportion of their monthly income spent on commuting expenses are more inclined to favor CB. Therefore, it can be observed that, at the present stage, CBs are more attractive to middle-aged women with middle to high incomes.

(2) The Influence of Psychological Latent Variables on Travel Mode Choice.

In the HCM, the coefficients of individual socioeconomic attributes change compared to the binary logit model. This indicates that the impact of individual

Table 4.5 Definition of utility variables

	Variable	Symbol	Variable type	Explanation
Individual attribute	Gender	GD	Dummy	Female (GD = 1), Male (GD = 0)
	Age (Reference: 40 years and older)	AG_1	Dummy	Age 19–30 ($AG_1 = 1$), otherwise ($AG_1 = 0$)
		AG_2		Age 31–40 ($AG_2 = 1$), otherwise ($AG_2 = 0$)
	Ownership of a private car	CR	Dummy	Owns private cars (CR = 1), otherwise (CR = 0)
	Monthly income (Reference: 10,000 yuan or more)	MI_1	Dummy	Less than 5000 yuan ($MI_1 =$ 1), otherwise ($MI_1 = 0$)
		MI_2		5001–10,000 yuan ($MI_2 =$ 1), otherwise ($MI_2 = 0$)
	Proportion of commuting expenses	DP	Continuous	Monthly commuting expenses as a proportion of monthly income
Attribute shared by travel mode	Waiting time	WT	Continuous	Minutes
	Travel time	TT	Continuous	Minutes
	Fare	TC	Continuous	Yuan
Attribute specific to PT	Availability of seats	ST	Dummy	Has a seat available (ST = 1), otherwise (ST = 0)
	Number of transfers	TF	Continuous	/
Attribute specific to CB	Detour ratio	DP	Continuous	Proportion of detour distance to total travel distance
	Discount rate	DC	Continuous	Fare discount, e.g.0.9: 10% discount
Latent variable	Comfort	η_1	Continuous	/
	Convenience	η_2	Continuous	/
	Reliability	η_3	Continuous	/

socioeconomic attributes on utility is manifested in two ways: directly in the utility function and indirectly through latent variables. The results show that reliability has the greatest impact, with an odds ratio of $\exp(0.086) = 1.09$. This means that for every one-level increase in the perceived reliability rating, the probability of choosing CB increases by a factor of 1.09. According to the measurement results of the reliability index in Table 4.3, the average perceived reliability for passengers at the present stage is 4.22. This translates to a positive utility equivalent to a reduction in travel time of 6.15 min, a reduction in waiting time of 4.97 min, or a fare discount of 2.34 yuan. It is a significant factor that cannot be ignored in travel mode choice. Since the majority of surveyed travelers are commuters, CB operators and governments

Table 4.6 Estimation results of model

		BLM		HCM	
		Coefficients	t-value	Coefficients	t-value
Constant		4.572***	5.541	19.366***	7.699
Gender (1: Female; 0: Male)		0.313***	2.578	0.525	-0.748
Age (19–30 years)		-0.462**	-2.073	-0.925**	-2.262
Age (31–40 years)		-0.515**	-2.380	-0.988**	-2.502
Ownership of a private car (1: Yes; 0: No)		-0.159*	-1.675	-1.582*	1.805
Monthly income (Less than 5000 yuan)		-0.612***	-3.146	-3.672***	-3.504
Monthly income (5001–10,000 yuan)		-0.167	-1.110	-1.888***	-2.834
Monthly commuting expenses as a ratio of monthly income		1.377***	4.326	3.626**	2.161
Comfort				0.030	0.061
Convenience				0.002	1.359
Reliability				0.086***	2.601
Travel time/min	Mean	-0.031***	-4.224	-0.059***	-3.848
	Standard Deviation	-	-	0.074***	5.420
Waiting time/min	Mean	-0.022**	-2.370	-0.073***	-2.902
	Standard Deviation	-	-	0.057**	2.120
Fare/yuan	Mean	-0.137***	-5.844	-0.155***	-3.262
	Standard Deviation	-	-	0.067	1.487
Seat availability (1: Yes; 0: No)	Mean	0.473***	4.991	0.703***	3.702
	Standard Deviation	-	-	1.928***	-6.888
Number of transfers	Mean	-0.018	-0.179	0.033	0.163
	Standard Deviation	-	-	0.321	1.308
Discount rate	Mean	-2.818***	-3.792	-13.761***	-7.533
	Standard Deviation	-	-	0.975***	-3.312
Detour ratio	Mean	-3.645***	-6.152	-10.737***	-8.021

(continued)

Table 4.6 (continued)

		BLM		HCM	
		Coefficients	t-value	Coefficients	t-value
	Standard Deviation	-	-	4.467***	11.275
Log-likelihood		-1640.4		-132.3	
McFadden R^2		0.043		0.229	

Note *** represents $p < 0.01$; ** represents $p < 0.05$; * represents $p < 0.1$

should focus on improving the reliability of operating times on various routes based on travelers' subjective perceptions to ensure continued choices by travelers. From the analysis above, it can be concluded that by constructing and measuring unobservable psychological factors through the MIMIC model, and then directly considering the role of latent variables in the behavioral choice model, rather than disregarding them as random errors, a better interpretation of the decision mechanism for travel mode choice can be achieved.

(3) The Impact of Travel Mode Attributes on Travel Mode Choice.

In the binary logit model, coefficient estimation results are for non-random terms, whereas in the HCM, except for the coefficient of the conventional bus transfer variable, all other travel mode attribute variable coefficients are significant. Additionally, different travelers exhibit normally distributed coefficients for travel time, waiting time, whether conventional buses have seats, CB detour rates, and discount variables in the utility function. The estimation results for travel attribute coefficients better describe the perceptual differences among different passengers regarding travel attributes. For example, whether conventional buses have seats. The variable's estimated mean is 0.703, with a standard deviation of 1.928. As shown in Fig. 4.3, the cumulative probability of coefficient estimates greater than 0 is 64.2%. This indicates that having seats in conventional buses increases the choice probability for 64.2% of travelers. It confirms that the primary reason travelers are attracted to CB is their offering of "one seat per passenger." Similarly, CB detours and fare discounts have a significant impact on the choice of 99% of travelers.

4.4.3 *Willingness to Pay (WTP)*

According to the theory of choice behavior, travelers' WTP for different travel attributes can be expressed as the ratio between the coefficients of travel mode attributes and price variables. This ratio reflects the additional cost that travelers are willing to incur for a unit change in a specific attribute at the unit price level. Based on the estimation results of the binary logit model, the average WTP for travel time and waiting time for travelers is calculated to be 13.58 yuan/hour and 9.64 yuan/hour, respectively. In contrast, the HCM yields average WTP values of 22.84 yuan/

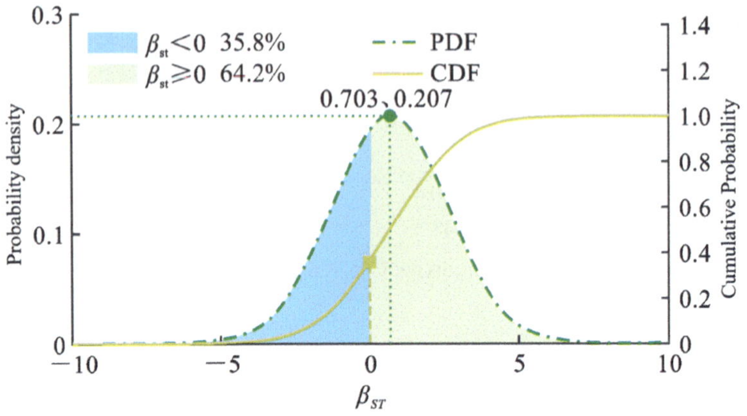

Fig. 4.3 Probability distribution of β_{ST}

hour and 28.26 yuan/hour, which align more closely with the average hourly wage levels of the respondents. Similarly, for travelers, the discount required for each 10% increase in detouring by CBs can be expressed as the ratio between the detouring rate and the discount rate coefficient:

$$dc_n = 0.1 \times \frac{\beta_{nDP}}{\beta_{nDC}} \tag{4.8}$$

Without considering individual differences among travelers, an 8% fare discount can offset the disutility caused by route detours. For instance, for a route covering a distance of 10 km with a fare of 10 yuan, if a CB adds 1 km of detour, the operating company would need to provide an average discount of 0.8 yuan. Assuming that the coefficient variables are mutually independent and keeping the discount coefficient variable constant, the required discount value for travelers due to route detours still follows a normal distribution $N \sim \left(0.078, 0.033^2\right)$, as shown in Fig. 4.4. In the same detour scenario, when accounting for individual heterogeneity, if the goal is to meet the needs of 90% of travelers, the discount rate should be 12% of the original fare. If no discount is provided, only 0.9% of passengers would remain.

4.5 Conclusions

Faced with various challenges in the operation of CB services in several cities across China, including low ridership, difficulty in balancing operational flexibility with stable traveler demand, this study aimed to explore the mechanism behind CB service choices by surveying CB passengers in Dalian. The study considered the diversity in travelers' needs and their underlying psychological factors. The main conclusions are as follows:

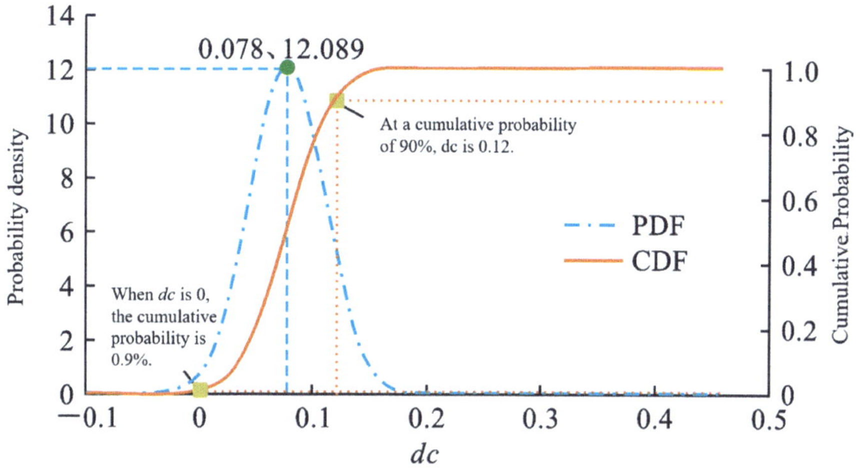

Fig. 4.4 Probability distribution of *dc*

Perceived Service Quality: Measurement indicators assessing passengers' perception of CB service comfort, convenience, and reliability confirmed the advantages of the CB operational model compared to conventional buses, private cars, and taxis. However, passengers still found issues with response times and service quality, and many mentioned long walking distances to CB stops, indicating a need for greater flexibility in route planning.

Socio-Economic Factors: Taking into account the influence of travelers' subjective attitudes, the MIMIC model revealed that socio-economic attributes significantly affected passengers' perceptions of CB service comfort, convenience, and reliability. The study found that high-income young adults had higher expectations for CB service quality and represented a significant potential user group. To better meet the needs of this demographic, CB services should focus on continuously improving reliability and other service quality aspects.

Mixed Logit Model Insights: By combining a latent variable framework with a flexible Mixed Logit model structure, the study revealed that attitude variables such as reliability had an equally important impact on travel choices as traditional service level variables like travel time. The model also captured unobservable individual heterogeneity. For instance, not having a seat on a CB reduced the choice probability for 64.2% of users. Dynamic adjustments to CB routes and stops, along with fare adjustments, should be considered to attract travelers continuously. Furthermore, taking into account the diversity in travel demands, for every 10% increase in detouring, at least a 12% fare discount is needed to ensure that 90% of travelers do not switch to other modes of transportation.

Despite these findings, the study has its limitations. The impact of latent variables on CB travel mode choices only considered subjective attitudes like comfort, convenience, and reliability. Future research should expand the scope of latent variables and explore the relationships among them. Additionally, understanding and catering to the differentiated demands of various user groups will be a major challenge in the future.

References

Anable, J. 2005. 'Complacent car addicts' or 'Aspiring environmentalists'? Identifying travel behaviour segments using attitude theory. *Transport Policy (Oxf)* 12: 65–78. https://doi.org/10.1016/J.TRANPOL.2004.11.004

Ben-Akiva, M., D. Mcfadden, K. Train, J. Walker, C. Bhat, M. Bierlaire, D. Bolduc, A. Boersch-Supan, D. Brownstone, D.S. Bunch, A. Daly, A. De Palma, D. Gopinath, A. Karlstrom, and M.A. Munizaga. 2002. Hybrid choice models: Progress and challenges. *Marketing Letters* 13: 163–175. https://doi.org/10.1023/A:1020254301302/METRICS.

Cao, Y., and J. Wang. 2016b. The key contributing factors of customized shuttle bus in rush hour: A case study in Harbin city. *Procedia Engineering* 137: 478–486. https://doi.org/10.1016/J.PROENG.2016.01.283

Chen, J., and M. Gan. 2014. A review of latent variable on urban travel behavior. *Journal of Transportation Systems Engineering and Information Technology* 14: 21–29. https://doi.org/10.16097/j.cnki.1009-6744.2014.06.004

Chen, X., Y. Wang, Z. Dai, and X. Ma. 2020. Research on demand identification for customized bus based on multi-source mobility data. *Big Data Research* 6: 105–118. https://doi.org/10.11959/j.issn.2096-0271.2020057

Cheng, L., X. Chen, and S. Yang. 2016. Low-income commuters' mode choice utilizing attitude-behavior model. *Journal of Transportation Systems Engineering and Information Technology* 16: 176–181+201. https://doi.org/10.16097/j.cnki.1009-6744.2016.01.027

Fu, X., Y. Gu, and Z. Liu. 2019. Scheduling activity and travel patterns in multi-modal transit networks with customized bus services. *Journal of Transportation Systems Engineering and Information Technology* 19: 20–27. https://doi.org/10.16097/j.cnki.1009-6744.2019.04.004

Hess, S., G. Spitz, M. Bradley, and M. Coogan. 2018. Analysis of mode choice for intercity travel: Application of a hybrid choice model to two distinct US corridors. *Transportation Research Part a: Policy and Practice* 116: 547–567. https://doi.org/10.1016/J.TRA.2018.05.019.

Hess, S., and K. Train. 2017. Correlation and scale in mixed logit models. *Journal of Choice Modelling* 23: 1–8. https://doi.org/10.1016/J.JOCM.2017.03.001.

Hörcher, D., D.J. Graham, and R.J. Anderson. 2017. Crowding cost estimation with large scale smart card and vehicle location data. *Transportation Research Part B: Methodological* 95: 105–125. https://doi.org/10.1016/J.TRB.2016.10.015.

Jian, Y., J. Zhangs, and song, Jin. 2019. Analysis of influencing factors of residents' willingness to custom bus travel. *Journal of Wuhan University of Technology (Transportation Science & Engineering)* 43: 247–252.

Jing, P., Z. Juan, and Q. Zha. 2014. Incorporating psychological latent variables into travel mode choice model. *China Journal of Highway and Transport* 27: 84–92+108. https://doi.org/10.19721/j.cnki.1001-7372.2014.11.012

Jing, P., J. Wang, Q. Zha, M. He, and Y. Lu. 2017. Influence of psychological factors on school travel mode choice a systematic review. *China Journal of Highway and Transport* 30: 107–119. https://doi.org/10.19721/j.cnki.1001-7372.2017.02.013

Ju, P., J. Zhou, H. Xu, and J. Zhang. 2017. Travelers' choice behavior of car sharing based on hybrid choice model. *Journal of Transportation Systems Engineering and Information Technology* 17: 7–13. https://doi.org/10.16097/j.cnki.1009-6744.2017.02.002

Li, Z., K. Liu, and J. Wang. 2021. Exploring obsrved and unobserved preference heterogeneity in choice behavior of demand responsive customized bus service. *Journal of Urban Planning and Development* 147: 1–7. https://doi.org/10.1061/(ASCE)

Paulssen, M., D. Temme, A. Vij, and J.L. Walker. 2014. Values, attitudes and travel behavior: A hierarchical latent variable mixed logit model of travel mode choice. *Transportation (Amst)* 41: 873–888. https://doi.org/10.1007/S11116-013-9504-3/TABLES/4.

Peduzzi, P., J. Concato, E. Kemper, T.R. Holford, and A.R. Feinstem. 1996. A simulation study of the number of events per variable in logistic regression analysis. *Journal of Clinical Epidemiology* 49: 1373–1379. https://doi.org/10.1016/S0895-4356(96)00236-3.

Qi, H., J. Xia, G. Wang, N. Jia, and Z. He. 2021. A behavioral intention to use model of autonomous vehicle ride-hailing incorporating traveler habit and altruistic preference. *Journal of Transportation Engineering and Information* 19: 1–10. https://doi.org/10.3969/j.issn.1672-4747.2021.02.001.

Raveau, S., M.F. Yáñez, J. de Ortúzar, and D.,. 2012. Practical and empirical identifiability of hybrid discrete choice models. *Transportation Research Part B: Methodological* 46: 1374–1383. https://doi.org/10.1016/J.TRB.2012.06.006.

Schlüter, J., A. Bossert, P. Rössy, and M. Kersting. 2021. Impact assessment of autonomous demand responsive transport as a link between urban and rural areas. *Research in Transportation Business & Management* 39: 100613. https://doi.org/10.1016/J.RTBM.2020.100613.

Vredin Johansson, M., T. Heldt, and P. Johansson. 2006. The effects of attitudes and personality traits on mode choice. *Transportation Research Part A Policy Practice* 40: 507–525. https://doi.org/10.1016/J.TRA.2005.09.001

Wang, J., Y. Cao, and Y. Wang. 2018. Customized bus route vehicle schedule method considering travel time windows. *China Journal of Highway and Transport* 31: 143–150.

Wang, J., T. Yamamoto, and K. Liu. 2019a. Role of customized bus services in the transportation system: Insight from actual performance. *Journal of Advanced Transportation* 2019: 6171532. https://doi.org/10.1155/2019/6171532

Wang, J., T. Yamamoto, K. Liu. 2019b. Role of customized bus services in the transportation system: Insight from actual performance. *Journal of Advanced Transportation* 2019. https://doi.org/10.1155/2019/6171532

Wang, J., T. Yamamoto, and K. Li. 2020a. Key determinants and heterogeneous frailties in passenger loyalty toward customized buses: An empirical investigation of the subscription termination hazard of users. *Transportation Research Part C Emerging Technologies* 115. https://doi.org/10.1016/j.trc.2020.102636

Yun, M., G. Liu, and F. Liu. 2017. Influence of change of public transportation service quality on travel mode choice behavior. *China Journal of Highway and Transport* 30: 119–125. https://doi.org/10.19721/j.cnki.1001-7372.2017.07.015

Zhang, Y., and P. Jiao. 2020. Analysis of customized bus demand-taking xi'an city as an example. *Journal of Beijing University of Civil Engineering* 36: 67–74.

Chapter 5
Continuous Choice Behavior of Customized Bus Users

Abstract Customized bus (CB) route planning and adjustment have an important impact on passenger occupancy rate and passenger's continuous choice behavior. Based on the stated preference survey on the continuous willingness to use of monthly-ticket CB users from 31 customized bus routes in Dalian, a binomial Logit model is constructed on the basis of random utility maximization theory. The result indicates that discount and detour distance are the two most significant influence factors of continuous willingness to use. The longer the travel distances are, the higher the willingness to continue using CB. The distance boundary with operational advantage between CB and conventional public transportation is 8 km in Dalian. For every additional 1.6 km of detour distance, a 10% fare reduction is required to maintain existing customized bus users, while at least 10% new subscribers should be induced to improve operation profit.

Keywords Customized bus · Demand response · Route adjustment · Detour distance · Stated preference survey

5.1 Introduction

Given the diversity and heterogeneity of individual travel needs, traditional PT services cannot fully meet the multifaceted demands of commuters, thereby prompting the emergence of CB services, a form of demand-responsive transit service. Since 2013, CB service has quietly risen in cities such as Beijing and Qingdao in China (Zhao et al. 2018). By offering seat reservations for every passenger and door-to-door transportation without the need for transfers, CB caters to the personalized travel preferences of riders. This demand-responsive bus service, based on new mobile internet technology, initially effectively attracted private car users to switch to customized buses upon its market entry (Liu and Ceder 2015), thereby alleviating urban traffic congestion to a certain extent (Zhang and Jiao 2020). Nonetheless, following years of rapid growth, passenger demand in some cities has abruptly declined due to various factors, leading to the cancellation of numerous CB routes.

J. Wang et al., *Customized Bus: User Travel Behavior and Demand Evolution*, Urban Sustainability, https://doi.org/10.1007/978-981-96-3488-0_5

The market demand evolution mechanism and service capability evaluation of CB have not received sufficient attention. "Demand-responsive travel" inevitably leads to dynamic evolution of demand and dynamic adjustment of operating routes, further leading to dynamic changes in service levels. This creates a persistent dilemma in customized bus operations: balancing route adjustments with demand stability.

5.2 Literature Review

Most relevant research focus on enhancing the operational efficiency of CB services from the perspectives of route optimization and vehicle scheduling. The route planning of CB aims to meet the route demands and improve operational efficiency. Existing studies have undertaken various attempts in terms of objectives, constraints, and algorithms. For example, a clustering algorithm based on station demand segmentation was proposed, grounded in considerations of passenger travel experience. From the perspective of spatiotemporal network modeling, ensuring the lowest occupancy rates and satisfying time window requirements were addressed (Huang et al. 2020). A three-stage hybrid coding approach was employed to develop a route optimization model catering to multiple depots, boarding/alighting stops, and customized buses (Wang and Ma 2019). Furthermore, multi-objective route optimization models that concurrently minimize operator costs and maximize passenger satisfaction have emerged as research hotspots (He Min et al. 2019). Compared to conventional PT services, CB scheduling necessitates greater consideration of travel time windows, stop constraints, bus capacity constraints (Wang et al. 2018), mixed-type vehicles constraints (Zheng et al. 2018), and more large-scale dynamic scheduling demands (Lei et al. 2017), aiming to elevate the demand service rate through real-time travel demand clustering.

Studies on passengers' willingness to choose CB services lay the foundation for forecasting future demand. Jian et al. (2019) analyzed the current operation status, service characteristics, and demand characteristics of CB and employed a structural equation model to study the impact of customized bus travel intention. Lai and Chen (2011) examined the interaction between satisfaction and travel intention, finding that passenger travel intention or loyalty significantly relies on passenger satisfaction. Wang et al. (2020) constructed four survival models of user loyalty by analyzing historical booking data of CB users, and comparing the differences in perceived service quality as an important reason for loyalty bias. Zhang et al. (2017) adopted an analytical approach combining demand analysis and network modeling to dissect the supply modes of CB services. Weng et al. (2017) compared the changes in passenger flow before and after the adjustment of fares from the perspective of the impact of fare adjustments on demand elasticity, and found sustained effects.

5.3 Research Area

Existing research has paid limited attention to the mutual influence between dynamic route adjustments and dynamic demand evolution. A profound understanding of users' continuous choice behavior can help transit operators improve services, which is crucial for retaining continuous users to the maximum extent. Currently, CB services in cities confront challenges such as neglected routes, low occupancy rates, and insufficient sign-ups to establish routes. If additional stops are chosen as a solution to address low occupancy rates, this will increase route detours, which in turn decrease CB users' travel satisfaction, potentially leading to further user attrition and forming a vicious cycle. Theoretically, there exists a balance between route detours and fare discount adjustments for CB service, aiming to maximize occupancy rates without reducing users' willingness to continue using the service. Therefore, this study, beyond the conventional decision factors of travel cost and time, further delves into the impacts of detour distance, travel distance, and fare discount mechanisms on the travel choices of existing CB monthly pass users. The objective is to provide a reference for dynamic route optimization, fare discount strategy formulation, and service level enhancement in customized bus operations.

5.4 Methodology Framework and Data Collection

5.4.1 Data Collection

Data for this study were collected through a survey questionnaire consisted of three parts. The first part focuses on travelers' personal information, including gender, age, monthly commuting expenses, number of vehicles owned by the household, routes taken, boarding and alighting stops, and other basic information. The second part involves a stated preference (SP) survey for travelers' mode choice behavior. It employed an orthogonal experimental design, where monthly pass users were presented with scenarios involving CB services or conventional bus + subway services. The third part examines users' perceptions of CB service quality, primarily covering three key indicators: comfort, convenience, and reliability, using a five-point Likert scale (Strongly Disagree, Disagree, Neutral, Agree, Strongly Agree).

Survey personnel conducted on-board surveys of 31 routes of Dalian Workers' Union Bus from July 3rd to July 5th, 2019, distributing questionnaires face-to-face to all monthly pass users on the buses.

In this survey, a total of 586 questionnaires were distributed, with each user allowed to respond only once. Questionnaires completed in a short time (less than 180s) or showing logical contradictions were considered invalid (140 in total), resulting in 446 valid questionnaires and an effective recovery rate of 76%. Among the surveyed monthly pass users, there were 75 male and 371 female respondents, with the majority falling within the age range of 19 to 40 years. The age group of 31 to 40 years old accounted for 56.3% of the sample. Each respondent answered the travel mode selection question six times, resulting in a total of 2676 valid samples.

The survey results on the service quality of CB services in Dalian are shown in Fig. 5.1. From the statistical results, users expressed overall satisfaction with the comfort, convenience, and reliability of CB service. Among the five indicators used to describe comfort, the assessment of smooth driving was the lowest. In terms of convenience, the vast majority of users were highly satisfied with the convenience of using the dedicated bus lanes during peak hours, while approximately half of the passengers expressed dissatisfaction with the customer service response of the CB service. Regarding reliability, the vast majority of passengers believed that notification services for vehicle arrival times should be provided.

5.4.2 Model and Utility Function

5.4.2.1 Binary Logit Model

This study investigates travelers' willingness to continue using CB services under the hypothetical scenario of dynamic adjustments to CB routes. The choice options include CB service and conventional PT service (bus + subway), where $i = 1$ represents CB service and $i = 2$ represents conventional PT service. The Logit model, chosen for its simplicity of structure and strong practicality, is a widely applied discrete choice model for analyzing multi-choice issues. The formula is as follows:

$$P_{1n} = \frac{e^{V_{1n}}}{e^{V_{1n}} + e^{V_{2n}}} = \frac{1}{1 + e^{-V_{1n} - V_{2n}}} \tag{5.1}$$

$$P_{2n} = 1 - P_{1n} = \frac{e^{V_{2n}}}{e^{V_{1n}} + e^{V_{2n}}} = \frac{1}{e^{V_{1n}} - e^{V_{2n}}} \tag{5.2}$$

where $P(i)$ represents the probability of choosing the $i - th$ ($i = 1{,}2$) travel mode; $V_i(X)$ represents the utility function for selecting the $i - th$ ($i = 1{,}2$) travel mode; n is the total number of travel modes (2 in this case).

The utility function $V_i(X)$ is as follows:

$$V_i(X) = U_i(X) + \varepsilon_i(X) \tag{5.3}$$

where $U_i(X)$ represents the fixed component of the utility function for the individual X choosing travel mode i; and $\varepsilon_i(X)$ represents the random component of the utility function for the individual X choosing travel mode i.

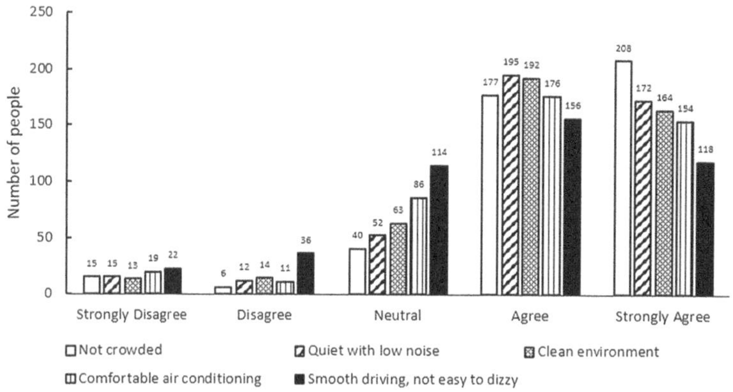

(a) comfort

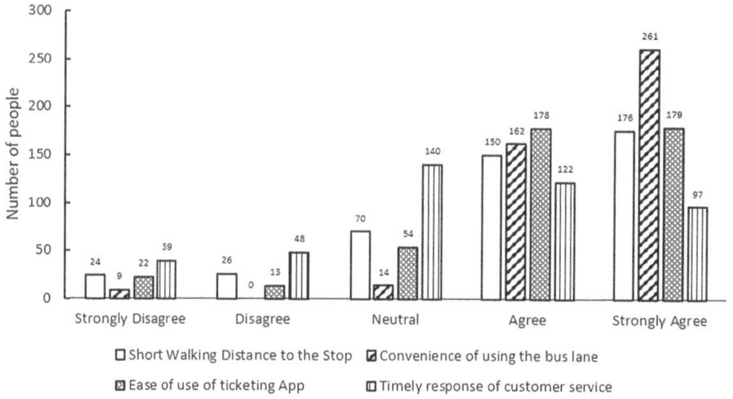

(b) convenience

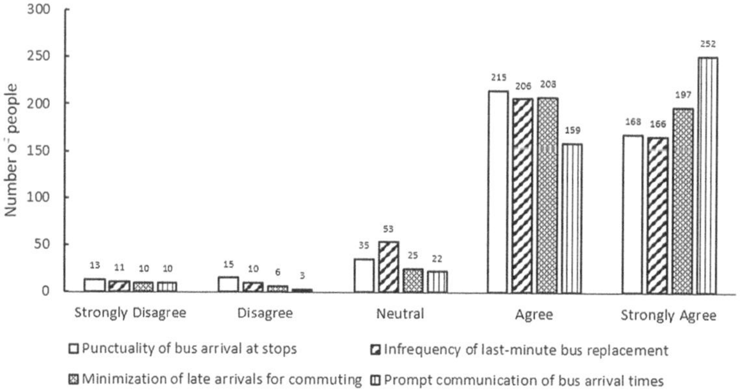

(c) reliability

Fig. 5.1 Perception distribution

Table 5.1 Utility variables and levels

Variable	Symbol	Levels
Travel time/min	X_1	10 levels for i=1: 30, 40, 45, 50, 60, 65, 70, 75, 80, 90 10 levels for i=2: 40, 45, 50, 60, 65, 70, 75, 80, 90, 100
Fare /yuan	X_2	4 levels for i=1: 2, 3, 4, 5 16 levels for i=2: 4.5, 6, 6.4, 7, 8, 8.8, 9, 10.8, 12, 12.8, 13, 13.5, 14.4, 15, 15.3, 18
Discount rate	X_3	3 levels for both: 100% (no discount), 90%, 80%
Transfer	X_4	2 levels for both: 1 (Yes), 0 (No)
Waiting time/min	X_5	3 levels for i=1: 3, 8, 15 3 levels for i=1: 5, 13, 20
Seat availability	X_6	2 levels for both: 1 (Yes), 0 (No)
Detour distance/km	X_7	3 levels for both: 0, 3, 5
Distance/km	X_8	3 levels for both: 10, 20, 25
Gender	X_9	2 levels for both: 1 (Male), 0 (Female)
Age	X_{10}	2 levels for both: 1 (above 30), 0 (19–30)
Ownership of a private car	X_{11}	2 levels for both: 1 (Yes), 0 (No)
Monthly income/yuan	X_{12}	2 levels for both: 1 (7500 and above), 0 (below 7500)
Monthly commuting expenses/yuan	X_{13}	3 levels for both: 150, 300, 400

Note $i = (1, 2)$, where i=1 represents CB service and $i = 2$ represents conventional PT service

5.4.2.2 Utility Influencing Factors

There are many factors that influence the actual travel choices of CB users. The variables of the choice options include common variables for both modes (travel time, fare, waiting time, travel distance), specific variables of PT (transfer and seat availability), and specific variables of CB (discount rate and detour distance). Among these, discount rate and detour distance take into account the needs of CB route adjustments and operational management, which are the key influencing factors studied in this paper. Personal variables of travelers include gender, age, number of vehicles owned by the household, personal monthly income, and monthly commuting expenses. The utility variables and their respective levels are defined in Table 5.1.

5.4.2.3 Utility Functions

The utility function expressions for CB service (U_{1n}) and regular PT service (U_{2n}) are as follows:

$$U_{1n} = V_{1n} + \varepsilon_1 = a_1 * X_1 + a_2 * X_2 + a_3 * X_3 + a_5 * X_5 + a_7 * X_7 + a_8 * X_8 \\ + b_1 * X_9 + b_2 * X_{10} + b_3 * X_{11} + b_4 * X_{12} + b_5 * X_{13} + b_6 + \varepsilon_1 \tag{5.4}$$

$$U_{2n} = V_{2n} + \varepsilon_2 = a_1 * X_1 + a_2 * X_2 + a_4 * X_4 + a_5 * X_5 + a_6 * X_6 + \varepsilon_2 \tag{5.5}$$

where $a_1, a_2, a_3, a_4, a_5, a_6, a_7, a_8$ are the parameters of transportation mode attributes; b_1, b_2, b_3, b_4, b_5 are the parameters of traveler personal attributes; b_6 represents the fixed utility of CB.

5.5 Results and Discussions

In this study, the *R* software was utilized to programmatically solve for the parameter values of the binary logit model, with the maximum likelihood estimation method employed for estimating the regression parameters. The results of the model parameter analysis are presented in Table 5.2. The goodness of fit of the model was first examined. The goodness of fit ρ^2 is similar in meaning to R^2 in regression analysis. The value of ρ^2 ranges from 0 to 1, with values closer to 1 indicating higher model accuracy. However, in practice, the goodness of fit for many discrete choice models typically ranges from 0.1 to 0.2. Therefore, a ρ^2 value of 0.130 signifies an acceptable fit for the logit model. To validate the parameters, the t-values in the third column of Table 5.2 represent the *T*-test values corresponding to the parameter estimates. The absolute value of the t-value is used to assess whether the corresponding attribute significantly influences travelers' choices. In general, if the absolute value of t is greater than 1.96, the attribute is considered to affect travelers' choices significantly.

The model results confirm the research hypothesis. Among all the parameters, detour distance and fare discounts exhibit the highest significance, far exceeding other influencing factors. Additionally, travel time, waiting time, seat availability, gender, and monthly commuting expenses significantly impact the willingness to choose CB services. On the other hand, fare, transfer, age, car ownership, and monthly income have relatively minor effects on the sustained use of CB. Therefore, careful consideration should be exercised in adopting route adjustments and fare discount strategies for the sustainable operation of CB service. The results indicate that users who are female, under 30 years old, without private vehicles, and with higher incomes are more inclined to opt for CB service consistently.

Based on further analysis of the model's parameter interpretations, the value of travel time is approximately 42 yuan per hour, which aligns with current practical scenarios. Assuming the current fare level of CB remains unchanged, the utility of using PT is higher for trips shorter than 8 km, whereas CB services offer relatively greater utility beyond 8 km. When the travel distance reaches 27 km, even without a price discount, a detour within 1 km in route adjustment will not impact the utility of using CB. If additional stops require an increased detour distance, for every additional 1.6 km of detour, a 10% discount (i.e., 90% of the fare) is needed to ensure that the

Table 5.2 Estimation results of model

	Variable	Coefficient	t-value
Attribute shared by travel mode	Travel time/min	−0.019*	−2.489*
	Fare /yuan	−0.027	−0.888
	Waiting time/min	−0.028**	−3.497**
Attribute specific to PT	Transfer (1: Yes)	−0.231	−1.090
	Seat availability (1: Yes)	0.275**	2.854**
Attribute specific to CB	Discount rate	−6.868**	−6.439**
	Detour distance/km	−0.438**	−7.892**
	Distance/km	0.016	1.348
	Gender (1: Male)	−0.259*	−2.27*
	Age (1: above 30)	−0.055	−0.614
	Ownership of a private car (1: Yes)	−0.073	−0.775
	Monthly income (1: 7500 and above)	0.067	0.744
	Monthly commuting expenses	0.003**	4.204**
	Constant	6.754**	6.111**
Goodness of Fit (ρ^2)		0.136	
Adjusted Goodness of Fit ($\overline{\rho}^2$)		0.136	

Note '*' $p < 0.05$, '**' $p < 0.01$

willingness of passengers to continue using CB is not affected. In other words, from the perspective of maximizing occupancy rates, an increase in detour distance by 1.6 km necessitates an increase in passengers by over 10% to maintain the profitability of the route operation unaffected.

5.6 Conclusions

The survey research on monthly pass users of CB reveals that, for the long-term users, fare price is not a significant variable, while travel time, though significant, exhibits a relatively low parameter value. Users are more significantly influenced by waiting time, detour distance, and fare discounts. Furthermore, passengers exhibit a higher willingness to continue using CB for longer-distance trips. In the current urban development state of Dalian, travel of about 8 km marks the utility boundary between CB and PT service.

According to the survey results, the detours caused by the addition of stops by CB companies significantly reduce the passengers' utility value, becoming the main reason for the loss of long-term users. Offering certain fare discounts will significantly enhance the utility value of choosing CB, thereby attracting and retaining users. The research results indicate that, in the case of CB routes reaching 27 km, an additional 1 km detour can be added during dynamic route adjustments. Therefore, CB operators should not detour without restrictions solely to improve occupancy rates. Only by offering appropriate fare discounts can they maintain their existing user base. The study suggests that for every additional 1.6 km of detour distance, a 10% fare reduction is recommended (i.e., a 90% discount).

Furthermore, from the perspective of travel satisfaction, to further improve the service quality of customized buses, operators should improve driving smoothness, customer service efficiency, and dynamic travel information services (such as timely notifications of vehicle arrival times), aiming to increase the long-term survival rate of CB passengers.

References

He, Min, M. Li, W. Shui, and H. Qian. 2019. Influence of reliability and comfort on responsive custom bus route design. *Journal of Highway and Transportation Research and Development* 36: 145–151.

Huang, D., Y. Gu, S. Wang, Z. Liu, and W. Zhang. 2020. A two-phase optimization model for the demand-responsive customized bus network design. *Transportation Research Part c: Emerging Technologies* 111: 1–21. https://doi.org/10.1016/J.TRC.2019.12.004.

Jian, Y., J. Zhang, and J.I.N. Song. 2019. Analysis of influencing factors of residents' willingness to custom bus travel. *Journal of Wuhan University of Technology (Transportation Science & Engineering)* 43: 247–252.

Lai, W.T., and C.F. Chen. 2011. Behavioral intentions of public transit passengers—The roles of service quality, perceived value, satisfaction and involvement. *Transport Policy (Oxford)* 18: 318–325. https://doi.org/10.1016/J.TRANPOL.2010.09.003.

Lei, Y., P. Lin, and K. Yao. 2017. The network scheduling model and its solution algorithm of internet customized shuttle bus. *Journal of Transportation Systems Engineering and Information Technology* 17: 157–163.

Liu, T., and A. Ceder. 2015. Analysis of a new public-transport-service concept: Customized bus in China. *Transport Policy (Oxford)* 39: 63–76. https://doi.org/10.1016/j.tranpol.2015.02.004.

Wang, C., and C. Ma. 2019. Optimization of parking lot and multi-vehicle route of customized buses based on genetic algorithm. *Journal of Transport Information and Safety* 37: 109–117+127.

Wang, J., Y. Cao, and Y. Wang. 2018. Customized bus route vehicle schedule method considering travel time windows. *China Journal of Highway and Transport* 31: 143–150.

Wang, J., T. Yamamoto, and K. Liu. 2020. Key determinants and heterogeneous frailties in passenger loyalty toward customized buses: An empirical investigation of the subscription termination hazard of users. *Transportation Research Part C: Emerging Technologies* 115. https://doi.org/10.1016/j.trc.2020.102636.

Weng, J., Q. Tu, P. Lin, and S. Xu. 2017. Impact analysis of public transport price adjustment on travel choice for travelers. *Journal of Wuhan University of Technology (Transportation Science & Engineering)* 41: 419–424+429.

Zhang, J., D.Z.W. Wang, and M. Meng. 2017. Analyzing customized bus service on a multimodal travel corridor: An analytical modeling approach. *Journal of Transportation Engineering, Part a: Systems* 143: 04017057. https://doi.org/10.1061/jtepbs.0000087.

Zhang, Y., and P. Jiao. 2020. Analysis of customized bus demand-taking Xi'an City as an example. *Journal of Beijing University of Civil Engineering* 36: 67–74.

Zhao, X., S. Zhang, G. Kong, and T. Jia. 2018. Discussion on the value of customized city bus to urban traffic control-taking the Shenzhen e-bus quality bus service for example. *Urban Public Transport* 42–46.

Zheng, H., X. Zhang, and Z. Wang. 2018. Design of demand-responsive service by mixed-type vehicles. *Journal of Transportation Systems Engineering and Information Technology* 18: 157–163.

Chapter 6
Heterogeneity in Travel Behavior Between Active Users and Inactive Users

Abstract Understanding the mechanism of continuous subscribing behavior is vital to the operation and even survival of CB systems. Capturing the differences in the subscribing behavior of active and inactive users and developing targeted and differentiated marketing strategies accordingly will help to improve the user retention rate. The results of this study revealed differences in the subscribing behavior between active and inactive users and the influence of a wide range of factors on the subscription behavior between the two groups, such as area-related and individual characteristics. Travel distance presents an inverted "U" shape effect on the propensity to use CB with the vertex at around 16 km for active users and 28 km for inactive users, which is a further finding of the monotonic positive effect indicated by the existing studies. For every RMB 10 cheaper CB trip, the likelihood that an active user holds a higher "propensity" to use CB services will increase by 8%. The cost advantage will attract novelty-seeking travelers to try CB services occasionally, but cannot convert them into high-frequency users. The passengers who started to subscribe to the CB service in winter are more likely to travel by CB frequently no matter in active and inactive user groups. The findings of this empirical study lay a theoretical foundation for ridership retention and marketing orientation of the customized and demand-responsive transit services.

Keywords Differential travel behavior · Active users · Inactive user · Continuous choice behavior

6.1 Introduction

As an innovative demand-responsive transportation (DRT) service, customized bus (CB) has been springing up in some countries in recent years (Abdullah 2017; Liu and Ceder 2015). Similar to other DRT systems, CBs collect travelers with similar travel demand, i.e., similar origin-destinations (ODs) as well as expected departure and arrival times, and aim to provide high-quality, point-to-point, customized transit services (Wang et al. 2019). User travel requests are put as a top priority throughout

J. Wang et al., *Customized Bus: User Travel Behavior and Demand Evolution*, Urban Sustainability, https://doi.org/10.1007/978-981-96-3488-0_6

the service design and operation. The CB routes and timetables are customized to the passengers recruited, and the systems, therefore, are suffering from higher operating costs in contrast to the regular bus transit service. In practice, only the routes with high occupancy rates, namely 80% total seat number of a bus, can be successfully launched, and only the launched routes with enough continuous subscriptions have been maintained. Failure to catch the trend of demand evolution leads to cancellation of already launched routes due to the low occupancy rates, making the high-cost operation overwhelmed with notable examples in China, including Guangzhou, Nanjing, Xiamen (Chu 2015; Ma 2015; Wang et al. 2019; Yan 2014). The flexible route design and demand-driven nature make the CB system extremely sensitive to the extreme volatility of demand (Currie and Fournier 2020). Therefore, investigating the subscribing behaviors of CB users is urgent and imperative for the operation and even the survival of CB systems.

The frequency with which customers repeat their purchases during a period has been frequently analyzed as a behavioral consequence of user loyalty toward a product or service (Aldas-Manzano et al. 2011; Drengner et al. 2010; García Gómez et al. 2006; Tranberg and Hansen 1986). In the field of transportation, the frequency of riding a mode of transportation is commonly investigated to reveal the mechanism of individual travel behavior towards a specific travel mode (Cao 2010; Cao et al. 2006; Yang et al. 2019) or namely their propensity to use the mode (Wang et al. 2015). To the best of our knowledge, existing research rarely investigated how different the travel behavior is between active passengers and inactive users. Obviously, the mechanism of continuous subscribing behavior of novelty-seeking and follow-the-leader users is different from that of loyal passengers. In the public transportation context, the cost of attracting and acquiring a new customer is generally considerably higher than that of retaining an existing one (Webb 2010). Capturing the differences in subscribing behavior between active users and inactive users and developing targeted and differentiated marketing strategies accordingly will help to improve the passenger retention rate, which is more important for DRT systems. Therefore, this paper pursues the objective of investigating the subscribing behavior of CB users and exploring the differences in effects of a wide range of factors between active and inactive passengers.

The factors in terms of service-related, area-related, and individual characteristics were extracted from the operational data of a successful CB system and additional open-access geospatial data sources. The active passengers and inactive users were distinguished, and differences in their subscribing behavior were investigated through a comparative analysis.

The remaining parts of this paper are structured as follows: Sect. 6.2 focuses on an extended review of studies pertaining to the travel behavior analysis of DRT users; Sect. 6.3 focuses on the methodology employed in this study; the research area, data collection, and variable selection are addressed in Sect. 6.4; finally, the results and discussions, and the conclusions drawn based on the key findings of this study are presented in Sect. 6.5 and Sect. 6.6, respectively.

6.2 Literature Review

DRTs have been widely applied all over the world during past decades and are seriously considered as a means of providing additional mobility at a lower overall cost in contrast to conventional public transportation services, albeit at a higher per passenger trip cost (Mulley and Nelson 2009; Nelson and Phonphitakchai 2012; Wang et al. 2014). Supported by the rapid development of information and communication technologies, the CB services, as one of its main developments, have been rapidly spread in major cities of China and some other countries in recent years (Abdullah 2017; Liu and Ceder 2015), holding the promise of establishing more attractive and diverse public transportation services and thereby reducing urban traffic congestion, especially during peak hours. Nevertheless, the DRT services have often been criticized owing to their relatively high cost of provision, their lack of flexibility in route planning, their inability to manage high demand, and uncertainty of traveling reliability (Currie and Fournier 2020; Mageean and Nelson 2003; Nelson and Phonphitakchai 2012). As a demand-driven transportation system, understanding and identifying the passengers' subscribing behavior becomes even more important to the system operation and even survival (Wang et al. 2021, 2020b).

The frequency with which customers repeat their purchase during a period has been frequently recognized as a behavioral consequence of user satisfaction or loyalty toward the product or service they purchase (Aldas-Manzano et al. 2011; Drengner et al. 2010; García Gómez et al. 2006). Plenty of literature employed the frequency of use toward a particular travel mode as the travel behavior and explored its relationship with internal and external factors, such as quality of service and built environment (Cao 2010; Kieu et al. 2018; Wang et al. 2015; Yang et al. 2019). Even though it is more important in DRT systems, existing studies have barely been involved in the long-term travel behavior in these systems, probably due to the challenges of data collection. In order to gain some foresight in variable selection and modeling, it is worthwhile to perform an extensive review of the key studies pertaining to the travel behavior studies of DRT systems and public transportation systems.

Wang et al. (2015) identified the factors that impact the use of DRT services into three categories: service related, area related, and individual-related factors. In terms of the service-related factors, Redman et al. (2013) summarized seven attributes to assess the service quality of public transportation systems, including reliability, frequency, price, speed, access, comfort, and convenience. Bass et al. (2011) analyzed the migration from public transportation to other modes based on survey data in Santiago and found that the passengers are more sensitive to the quality of services than fare changes. Li et al. (2013) examined the service reliability, dwell time (buses' stop time at a station), and running speed of the Beijing public transportation system and identified the running speed and service reliability as the key factors of service quality. Quadrifoglio et al. (2006) evaluated the performance of a DRT system and pointed out that the velocity should remain above a minimum threshold to attract and retain users. Additionally, DRT routes for long-distance trips are more attractive, and paying more attention to distant trips is identified as one of the key characteristics of successful DRT services (Kirby and Bhatt 1975; Zhang et al. 2017). Through a numerical analysis based on an idealized multimodal travel corridor, Zhang et al. (2017) identified three key factors that significantly motivate the mode shift from others to CBs. They are fare, usage of bus lanes, and out-of-vehicle travel time. Waiting time, cleanliness, and comfort are also important factors for service quality of those services (Dell'Olio et al. 2011; Lierop et al. 2018), in addition to the travel attributes in terms of cost, time, and distance of trips, which were most concerned in the studies discussed above and generally used to catch pictures of service quality.

Traditional public transit services are known to work efficiently in densely populated areas, while flexible-route transit services such as DRT services are taken as a cure for high-cost bus service in low-demand areas (Nourbakhsh and Ouyang 2012; Scott 2010). Based on a case study of a DRT system in rural areas of Lincolnshire, Wang et al. (2015) indicated that a unit increase in population density would decrease the frequency of DRT trips by 1.3% on average. DRT service is more of a deep rural product than an urban solution in Great Britain (Davison et al. 2014). The areas with more elderly, mobility limited, and low-income residents aggregated are protentional markets of DRT systems in North America (Transit Cooperative Research Program 2004). Additionally, Wang et al. (2014) modeled the demand of DRT in Great Manchester with supply-oriented factors and socio-economic factors and revealed that the high demand was associated with low car ownership level, a higher proportion of white people, and high levels of social deprivation. They found that the area-related factors account for about 20% variation of DRT demand in the model. Other factors, including land-use patterns, existing transportation accessibility, and access to other alternatives, were also found to affect the travel behavior of residents significantly (Cao et al. 2007; Tao et al. 2020) but have not yet been paid enough attention to the behavior analysis of DRT uses.

As area-related factors indicated above, socioeconomic characteristics such as age, gender, vehicle ownership, and income level have a direct influence on the use of DRT services. In particular, females are found as the dominant users of DRT services in Europe, owing to their poor access to a car (Mageean and Nelson 2003). In those systems, more than 65% of passengers are retired, stay-at-home persons, and students. Additionally, low workforce status, lack of train station proximity, longer waiting time, and longer walking time are identified to positively affect the usability of the services (Jain et al. 2017).

Moreover, the characteristics of user historical subscribing behaviors are indispensable in the forecast of user retention, e.g., the time interval between two sequential subscriptions, the time when the user first subscribed to the service, etc. (Allenby et al. 1999; Wang et al. 2020b). The perceived quality of service (satisfaction) for the first purchase (Chi 2012) and ride experiences with extreme weather (hot or cold weather makes a difference in riders' perception of waiting and transfer times) (Balcombe et al. 2004) also significantly affects the use of public transportation services.

However, most of the aforementioned studies investigated travel behavior based on the data derived from the stated preference and revealed preference surveys. The results obtained based on these types of conventional survey data are susceptible to the distributions of sampling, way of survey, design of questionnaire, and respondent-related attributes, which may lead to biased estimates of parameters, for instance, with the under-reporting issue (Yamamoto et al. 2008). Additionally, due to the high costs of the survey, it is difficult to follow all respondents over time to capture their long-term choice behaviors. The continuous subscribing behaviors of passengers were barely investigated. Fortunately, the operational data, such as the actual re-use observation, is helpful to overcome these limitations (can also be seen as revealed preference data) and is much more effective in capturing the long-term subscribing behavior of individuals. Additionally, a set of data mining techniques, including map-based route planning (Liu et al. 2016; Wang et al. 2019) and land use classification based on points of interest (POI) data (Hu and Han 2019; Yao et al. 2017), can be employed to derive and match information with respect to the service-related, area-related, and individual-level characteristics for each passenger.

6.3 Methodology

6.3.1 Identification of Active Users and Inactive Users

A user is identified as an active user if he/she has not terminated their subscriptions, otherwise, he/she is classified as an inactive user (Wang et al. 2020b). Based on the long-term observations on all registered users of the CB system in Dalian, China, we analyzed the long-term subscribing behavior of these passengers and summarized three subscription patterns that can be used to clarify whether passengers have terminated their subscriptions. In particular, a subscriber is identified as an active user in the following cases (subscription patterns) (as illustrated in Fig. 6.1):

- Case 1: a user was observed to subscribe to the service in the last month of the observation period;
- Case 2: a passenger temporarily stops his/her subscription at the end of the observation period, but the length of the suspension period is less than the maximum length of suspension in their historical subscription records;
- Case 3: a passenger temporarily stops his/her subscription at the end of the observation period for the first time, but the length of the suspension period is less than three months (derived based on the distribution of suspension time for all registered users).

Passengers who do not meet the above cases are identified as inactive passengers. We finally employed additional data from October 2017 to validate the identification results.

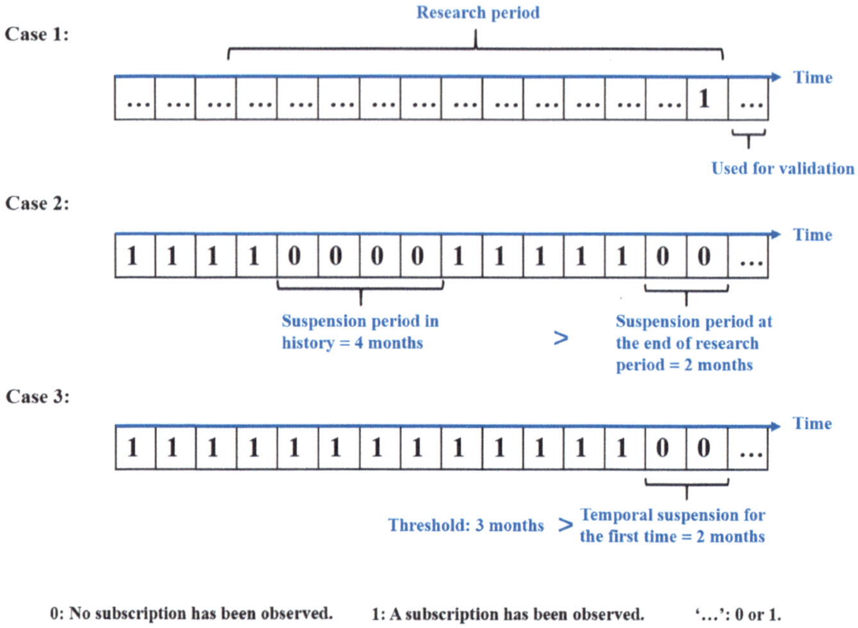

Fig. 6.1 An illustration of active user identification

6.3.2 *Subscribing Behavior Modeling*

Travel behavior as the dependent variable in this study is the observed subscription frequency within a research period. The travel frequency is naturally ordered in conceptual and econometric properties, and therefore the ordered choice model is commonly used as the analysis framework (Greene and Hensher 2010; Wang et al. 2015). Since all the users investigated in this paper are monthly subscribers (they only need to subscribe to the services once per month, and the subscription is valid for all the weekdays in that month), the largest frequency of use for one year is twelve times. The difference in propensity (to subscribe to the CB services) cannot be captured properly when using the original counts as categories, i.e., from 0–12, and some categories suffer from small sample sizes. It is often thought best to collapse categories containing small samples in order to improve the asymptotic approximations and fit of the model, particularly in situations where the data apparently deviate from the proportional odds assumption (Greene and Hensher 2010; Murad et al. 2003; Wang et al. 2015). Therefore, the observed subscription frequency was coded into 3 categories to indicate user "propensity" to subscribe to the CB services.

In particular, two criteria were used to collapse categories of the subscription frequency: (1) the categories generated are able to effectively distinguish the differences in the "propensity" (in which the model will have the largest log-likelihood at constant), and (2) the new categories generated will sacrifice least information in data (controlled by choosing the model with minimal Akaike Information Criterion value). Finally, the categories for active and inactive users are determined as follows (Table 6.1 lists the sample distribution of the finally determined categories):

as for active users

1 = infrequently (less than 7 times)
2 = ordinarily (8 to 10 times)
3 = frequently (11 to 12 times)

as for inactive users

1 = infrequently (no subscription observed)
2 = ordinarily (1 to 5 times)
3 = frequently (6 to 9 times)

The frequency of use observed within an observation period is always seen as the "propensity" to use the travel mode (Wang et al. 2015), and the "propensity" is potentially ordinal (the relative ordering of the potential subscribing patterns is known but the exact distance between them is not [Williams 2016]). Thus, the ordinal response model is a natural choice of modeling. The most popularly used model is the parallel-lines model (ordered logit/probit model), assuming that the effects of factors remain across the categories. However, this assumption is frequently violated in practice (Long and Freese 2014). The Brant test (Brant 1990) is commonly employed to test whether this assumption holds. As in our case, the total Brant test for active users and inactive users indicates the evident violation of the parallel-lines assumption. Therefore, we finally employed a generalized ordered logit model which can relax the parallel-lines assumption to avoid bias estimation.

Suppose, there is an unobserved, continuous, latent variable y_i^*, , which classifies the frequency of use into "propensity" categories (or namely the propensity to use the CB service) ranging from $-\infty$ to $+\infty$. The general model is determined by:

$$y_i^* = X_i\beta_j + \varepsilon_i \tag{6.1}$$

where j is a "propensity" category, $\boldsymbol{X}_i$ is a vector of observed explanatory variables for the *i*-th individual, β_j is a vector of parameters for the *j-th* category to be estimated,

Table 6.1 Sample distribution of categories

	Infrequently (1)	Ordinarily (2)	Frequently (3)	Total
Inactive users	2036	723	322	3081
	(66.08%)	(23.47%)	(10.45%)	(100%)
Active users	446	573	1097	2116
	(21.08%)	(27.08%)	(51.84%)	(100%)

Note Percentage of samples in parentheses

and ε_i is the error term that is assumed to be distributed logistically for the logit model.

For an ordinal outcome variable with J categories, the generalized ordered logit model in which the probability of getting a higher propensity to use CB services for a given subscribing frequency can be written as (Williams 2016):

$$P(y_i > j) = g(X_i\beta_j) = \frac{\exp(X_i\beta_j - \alpha_j)}{1+\exp(X_i\beta_j - \alpha_j)} j = 1, 2, \ldots, J - 1 \tag{6.2}$$

$$g(X_i\beta_j) = 1 - F(X_i\beta_j) \tag{6.3}$$

where the y_i is the observed frequency of CB use, α_j is the intercept, and $F(.)$ is the cumulative distribution function of ε. When the J is larger than 2, the generalized ordered logit model can be seen as a series of binary logistic regressions where categories of the outcome variable are combined. For instance, if $J = 3$, then $j = 1$ indicates that category 1 is contrasted with categories 2 and 3; for $j = 2$, categories 1 and 2 are contrasted with category 3. Therefore, a positive estimate of the coefficient β suggests that higher values of the independent variable increase the probability of a given user being in a higher category of the outcome variable compared to their current category. Conversely, a negative coefficient implies that higher values of the independent variable raise the likelihood of the user remaining in the current or a lower category of the outcome variable (Eluru and Yasmin 2015; Williams 2006).

According to Eq. (6.2), the probabilities that y_i will take on each of the values $1, 2, \ldots, J$ can be calculated as:

$$\begin{cases} P(y_i = 1) = 1 - g(X_i\beta_1) \\ P(y_i = j) = g(X_i\beta_{j-1}) - g(X_i\beta_j)\ j = 2, \ldots J - 1 \\ P(y_i = J) = g(X_i\beta_{J-1}) \end{cases} \tag{6.4}$$

In our case that the outcome variable has three possible values, the generalized ordered logit model will have two sets of estimated coefficients. The ordered logit model can be seen as a special case of the generalized model where the coefficients are the same for each j.

$$P(y_i > j) = \frac{\exp(X_i\beta - \alpha_j)}{1+\exp(X_i\beta - \alpha_j)} j = 1, 2, \ldots, J - 1 \tag{6.5}$$

where the α_j directly indicates the cut point in the order logit model. In between these two extremes is the partial proportional odds model (Williams 2016), in which some of the coefficients are the same while others can differ. With a partial proportional odds model, we can relax the parallel-lines assumption only for the variables in which the assumption is violated. Thus the Eq. (6.2) can be rewritten as:

$$(y_i > j) = g(X_i\beta_j) = \frac{\exp(X_{1i}\beta_{1j} + X_{2i}\beta_2 - \alpha_j)}{1+\exp(X_{1i}\beta_{1j} + X_{2i}\beta_2 - \alpha_j)} j = 1, 2, \ldots, J - 1 \tag{6.6}$$

where β_{1j} is a vector of coefficients associated with X_{1i} and varies across categories, and β_2 is a vector of coefficients associated with X_{2i} (the subset of variables for which the parallel-line assumption is not violated). The cut points are the function of independent variables in which their coefficients vary across categories (Eluru and Yasmin 2015), i.e., $cutpoint_j = X_{1i}\beta_{1j} - \alpha_j$.

Therefore, the log-likelihood function for a sample consisting of n observations with a constant variance can be written as:

$$LL(\boldsymbol{\beta}) = \sum_{i=1}^{n}\sum_{j=1}^{J} I_i \ln P(y_i = j) \quad (6.7)$$

where I is the indicator variable that is equal to 1 if the observed frequency of the *i-th* user results in the category j and 0 otherwise.

6.4 Data Collection and Modeling

6.4.1 Case Study Area

The study area chosen for this work is Dalian, which is one of the typical cities in China facing escalating traffic problems resulting from the drastically increasing car ownership in the past decade. Dalian has a population of more than 6.69 million (National Bureau of Statistics of the People's Republic of China 2011). The level of motor vehicle ownership reached 1.37 million at the end of 2016 (Dalian Municipal Bureau of Statistics 2017). The Dalian urban railway system is not well developed and consists of two subway lines and two rapid railway lines. The subway lines mainly provide transport services between main city centers, while the rapid railway lines connect the emerging city centers to the main city areas.

The first CB route in Dalian was launched in March 2014 to serve white-collar commuters traveling between some residential areas and several central business districts. After more than one year of test runs, the Dalian CB system was officially founded in August 2015 and started to collect demand from almost all urban areas. The Dalian CB system is regarded as one of the most successful CB systems in China (Liu and Ceder 2015; Wang et al. 2019).

6.4.2 Data Collection

The operational data from August 2015 to September 2017 of the Dalian CB system were acquired from the service operator, which plays a leading role in initiating the CB services in Dalian. As an emerging and demand-driven mode of transportation, it is necessary to cultivate its unique market at the earliest stage. After more than one year's trial run (started in March 2014), the number of passengers first reached more than 2500 in December 2015 and remained relatively steady thereafter. However, there is no evidence showing that any aggressive marketing strategy resulted in a spike in new users. After that, the retained users turned out to be the main component of passengers, and only a small part of new passengers were generated. Therefore, understanding the mechanism of continuous subscribing behavior of users is vital to the operation of CB services.

Figure 6.2 illustrates the changes in the number of users investigated in this study (the enrolled users who do not have a subscription in a given month are not counted in that month). Bars with the same color indicate that they subscribed to CB services for the first time ("born") in the same month.

This paper extracted the latest twelve months of obtained subscription records as the research period, i.e. from October 2016 to September 2017. However, the mechanism underlying the low frequency of use among passengers who subscribed to the services during the research period may differ from that of users who subscribed prior to the research period's inception. Thus, we only concentrated on the passengers who had started subscribing to the services before October 2016, as shown in Fig. 6.2. In total, 5197 passengers started to subscribe to the service before October 2016. We model the frequency of active users and inactive passengers separately, in which the differences in effects of the factors are focused on in this study.

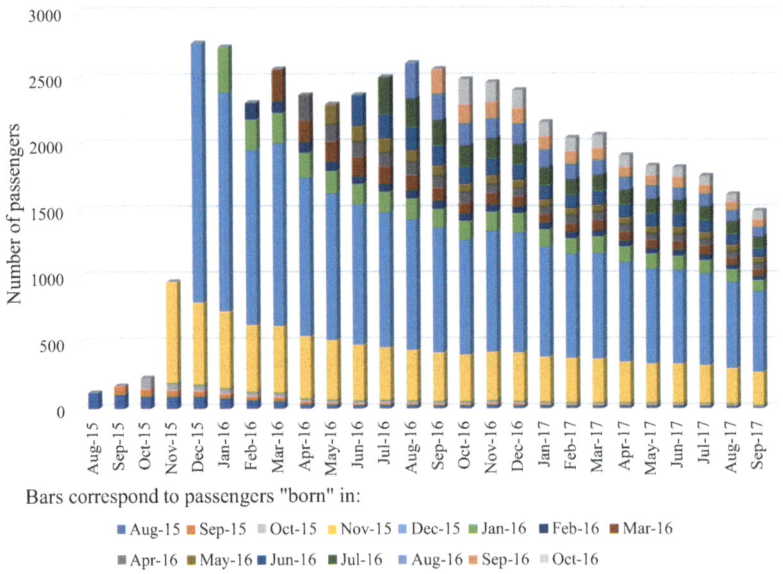

Fig. 6.2 Changes in the number of users during the observation period used in this study (n = 5197)

As shown in Table 6.1, 3081 out of the 5197 passengers have been identified as inactive passengers in the CB system, while the other 2116 users are active passengers. In particular, each travel frequency group (e.g., the group that users use CB frequently) of active and inactive users (as shown in Table 6.1) contains more than 10% of their entire sample sizes, indicating enough sample size for model estimation.

6.4.3 Data Description and Variable Selection

Travel attributes such as travel distance and travel expense are frequently employed as indicators of quality of service in public transportation services as reviewed in Sect. 6.2. The CB travel distance for each passenger was calculated through the route-planning program based on Baidu map. For more details about the route-planning program and settings to obtain these attributes, please refer to (Wang et al. 2019). Additionally, the travel expense by taxi was also obtained and the differential travel expense between taxi and CB was incorporated as a proxy of service quality. Please note that the financial support from the employers for commuting is not considered in this study due to the lack of related information.

In terms of the area-related factors, GDP, population density, and land-use patterns were frequently used in previous studies to model travel behavior (Balcombe et al. 2004; Giuliano and Narayan 2003). Fortunately, Huang et al. (2014) and Fu et al. (2014) generously shared their processed 1-km-scale population and GDP datasets based on the latest population census of the People's Republic of China conducted in 2010. The information in terms of population density and GDP was linked and matched for each CB passenger based on spatial location.

We mined the land-use pattern for the spatial units where a subscriber resides from accessible geospatial big data sources, such as POIs (Hu and Han 2019; Yao et al. 2017). In total, the land use pattern was finally classified into six primary types, including residences, administration and public services, commercial and business facilities, industrial and manufacturing spaces, roads and transportation, and green spaces and parks (Ministry of Housing and Urban–Rural Development of the People's Republic of China (MOHURD) 2012). The ratio of POIs for each land use category (Hu and Han 2019) was calculated, and the effects were examined in this study.

Previous studies have found significant heterogeneity of personal socioeconomic characteristics on the travel behavior toward demand responsive transit services (Davison et al. 2012; Loo et al. 2015; Wang et al. 2020b). Nevertheless, only the gender information is collected from the Dalian CB system due to privacy issues.

The experience of the first purchase is more important than the repeat (Chi 2012) and is always emphasized in the investigation of passenger loyalty (Wang et al. 2020a, 2020b). As an emerging service of transportation, the perceptible service quality in terms of convenience and efficiency in subscribing/returning tickets and information availability has been developed with the progress of the CB system. Therefore, the time when a user subscribes to the service for the first time (indicated by the "first subscription time") was extracted from the subscription records and incorporated into the models to explore this heterogeneity of service quality in the progress of the system. In particular, the first subscription time was coded as the month sequence order when a user subscribed to the service for the first time. June 2015 was set as the baseline and coded as 0.

DRT services have a shorter waiting time than conventional public transportation services and therefore are more attractive in extremely hot and cold climates (Balcombe et al. 2004). The cold winter in Dalian may contribute to attracting passengers from conventional public transportation. A winter dummy variable was therefore calculated and examined to determine whether a passenger started subscribing to the CB services in the winter. In particular, the "winter effect" can be captured with the winter dummy variable after controlling the month effect by incorporating the variable defined as the first subscription time.

Additionally, the order of boarding station (the order position of the boarding station for a passenger on his/her CB route.) also influences passenger loyalty (Wang et al. 2020b). For instance, the passenger who gets on a CB at a station with a small order position on the CB route is more likely to suffer from longer detours and therefore have a lower level of satisfaction.

The methodologies discussed in Sect. 6.3 were employed in this paper to model the subscribing frequency of CB users. The selection of variables was based on the z-statistics of coefficient estimates. Variables with quite small z-statistics were excluded from the models, while variables with at least one estimate at the 0.10 significance level were maintained in the final models. The descriptive statistics of variables were finally included in the models and their corresponding descriptions were summarized in Table 6.2.

6.5 Results and Discussions

Firstly, we estimated the ordered logit model and conducted the Brant test to check whether the parallel-lines assumption holds by using "ologit" and "brant" packages in STATA 15.1. The null hypothesis is that all the coefficients remain across categories of the outcome variable. In the current case, the parallel-lines assumption is violated for some variables, including travel distance, population density, first subscription time, and order of boarding station for the inactive user model, and female dummy and first subscription time for the active user model, while the others are not. The results illustrate that the partial proportional odds model is preferred for the dataset in this study (Table 6.3).

Table 6.4 presents the estimation results of the partial proportional odds models (using "gologit2" package in STATA 15.1) for inactive users and active users respectively (variables with the same parameters between categories have been bolded). Two streamlined models were also estimated to test the robustness of the model specifications. The estimation results are summarized in Table A1 in Appendix. The model specifications are tested for robustness, given that the sign and magnitude of

Table 6.2 Descriptive analysis of explanatory variables included in the models

Variable (Symbol)	Description	Inactive users				Active users			
Service-related attributes		Mean	Std. Dev	Min	Max	Mean	Std. Dev	Min	Max
Travel distance by CB (TD_{CB})	Travel distance by riding CB (10 km)	1.57	0.96	0.04	5.82	1.63	0.98	0.33	5.48
Differential travel expense between CB and taxi ($TE_{\text{taxi-CB}}$)	Difference in travel expenses (TE) between taxi and CB: $TE_{taxi-CB} = TE_{taxi} - TE_{CB}$ (10 RMB)	2.04	1.35	0.50	12.28	2.26	1.70	0.62	12.84
Area-related characteristics									
GDP per capita at origin	GDP per capita for the 1 km × 1 km unit cell around the origin of a user, calculated based on the economic census conducted in 2010 (10,000 RMB)	0.28	0.11	0	0.80	0.28	0.11	0	0.80
Ratio of residential-area POIs around the origin cell	Category ratio of residential-area POIs (%) in the 1 km × 1 km unit cell around the origin of a user	0.33	0.23	0	0.93	0.32	0.22	0	0.93
Population density at origin	Population per square kilometer around the origin of a user, calculated based on the population census conducted in 2010 (10,000 people)	0.74	0.70	0	2.35	0.71	0.67	0	2.35
Individual characteristics									
Gender (Female)	It is equal to 1 for female passengers; else, it is 0	0.74	0.44	0	1	0.79	0.41	0	1

(continued)

Table 6.2 (continued)

Variable (Symbol)	Description	Inactive users				Active users			
Service-related attributes		Mean	Std. Dev	Min	Max	Mean	Std. Dev	Min	Max
First subscription time	The order of month sequence when a user subscribes to the monthly subscription for the first time (June 2015 is taken as the baseline)	5.94	3.60	0	14	6.05	3.55	0	14
Order of boarding station	Order position of the boarding station for a passenger on his/her CB route	3.19	3.00	1	16	3.37	2.98	1	16
First subscription in winter	It is equal to 1 when a passenger started to subscribe to CB service in winter (Dec., Jan., and Feb.); else, it is 0	0.45	0.50	0	1	0.48	0.50	0	1

Table 6.3 Brant test of the parallel-lines assumption

	Inactive users		Active users	
	chi2	$p >$ chi2	chi2	$p >$ chi2
Overall	87.76	< 0.001	16.85	0.078
Travel distance by CB (TD_{CB})	13.23	< 0.001	0.70	0.403
Square of travel distance (TD_{CB}^2)	8.08	0.004	1.10	0.294
Differential travel expense between CB and taxi ($TE_{\text{taxi-CB}}$)	3.01	0.083	0.32	0.569
GDP per capita at origin	0.88	0.349	0.02	0.895
Ratio of residential-area POIs around the origin cell	1.52	0.217	0.60	0.439
Population density at origin	4.71	0.030	0.44	0.508
Female	0.30	0.586	6.78	0.009
First subscription time	42.76	< 0.001	3.27	0.071
Order of boarding station	4.05	0.044	0.48	0.488
First subscription in winter	0.27	0.601	0.15	0.698

Note A significant test statistic indicates that the parallel-lines assumption is violated variables are still preserved. Table 6.4 is focused in this Section to better interpret the effects of a wider range of factors.

Odds ratio (*OR*) and the percentage change in odds for a one-unit increase of each factor were calculated to assist interpretation:

$$OR_{x_k} = \exp(\beta_{jk}) \tag{6.8}$$

$$\Delta OR_{x_k} = (OR_{x_k} - 1) \times 100\% \tag{6.9}$$

where Eqs. (6.8) and (6.9) calculate the odds ratio and percentage change in odds for the *k-th* factor x_k respectively.

Figures 6.3 and 6.4 depict the relationship between travel distance and the probability of each category for female passengers with their first subscription in winter, with all other variables held at their mean values for both the active and inactive groups, respectively. The negative and significant coefficients of CB travel distance indicate an inverted "U" shape effect on the propensity to use CB service (see the curves for outcome = 3 in Fig. 6.3 and Fig. 6.4). The likelihood that an active passenger will subscribe to the CB service more frequently increases with his/her travel distance when it is less than about 16 km, while the threshold of travel distance for the inactive user is larger, i.e., 28 km. This finding is partially consistent with previous studies which indicated that the passengers with longer travel distances are more likely to subscribe to demand responsive transit services (Kirby and Bhatt 1975; Zhang et al. 2017), but they failed to identify a threshold point in the effects of travel distance. Although the CB services are more attractive for long-distance travel, the excessively long route distance (longer than 28 km) is not conducive to user retention.

As expected, the coefficient of differential travel expense between taxis and CBs shows positive effects on the CB subscribing frequency in the active user group, indicating that for every RMB 10 cheaper CB trip in contrast to the taxi's, the likelihood that a passenger subscribes to the service frequently will increase by 8%. Although the estimates are not very statistically significant (e.g., at the 90% confidence level), they can also provide us with some implications of the most probable direction and value of their effects. However, for the inactive user group, the effects of subscription price are inconsistent among the categories. For the panel outcome = 1 v.s. outcome = 2, 3, the effect of price is positive, while for the panel outcome = 1, 2 v.s. outcome = 3, the probability of getting a higher propensity to use the CB services decreases with the increase of the deferential expense. It indicates that the price advantage relative to the taxi services will attract novelty-seeking travelers to try CB services more frequently, but cannot convert them into high-frequency users. There must be some reasons that lead to this seemingly counterintuitive phenomenon, and we shall leave it as an important research topic for our future work.

The six land-use types were examined in the developed models, and only the ratio of residential area has been found to influence the CB subscribing behavior significantly. All included area-related factors show distinct effects on the "propensity"

Table 6.4 Estimation results of the partial proportional odds models

	Model 1 (Inactive passengers)				Model 2 (Active passengers)			
	Coefficient	z	Odds ratio	Percentage change in odds (%)	Coefficient	z	Odds ratio	Percentage change in odds (%)
outcome = 1 v.s. outcome = 2, 3								
Travel distance by CB (TD_{CB})	0.391**	2.42	1.48	47.82	**0.321****	**2.02**	**1.38**	**37.91**
Square of travel distance (TD_{CB}^2)	−0.081**	−2.35	0.92	−7.76	**−0.106*****	**−3.10**	**0.90**	**−10.08**
Differential travel expense between CB and taxi ($TE_{\text{taxi-CB}}$)	0.013	0.27	1.01	1.27	**0.077***	**1.73**	**1.08**	**8.01**
GDP per capita at origin	**−0.742****	**−2.12**	**0.48**	**−52.40**	**0.111**	**0.29**	**1.12**	**11.73**
Ratio of residential−area POIs around the origin cell	**0.242**	**1.13**	**1.27**	**27.34**	**−0.602****	**−2.54**	**0.55**	**−45.22**
Population density at origin	**0.144****	**2.26**	**1.16**	**15.54**	**−0.110**	**−1.52**	**0.90**	**−10.44**
Female	**0.084**	**0.93**	**1.09**	**8.74**	−0.348**	−2.44	0.71	−29.36
First subscription time	0.206***	15.39	1.23	22.84	0.051**	2.92	1.05	5.21
Order of boarding station	0.020	1.48	1.02	2.05	**0.026***	**1.71**	**1.03**	**2.59**
First subscription in winter	**0.192****	**1.98**	**1.21**	**21.11**	**0.126**	**1.27**	**1.13**	**13.39**
Constant	−2.512***	−9.00			1.092***	3.42		
Outcome = 1, 2 v.s. outcome = 3								
Travel distance by CB (TD_{CB})	1.245***	4.68	3.47	247.35	**0.321****	**2.02**	**1.38**	**37.91**
Square of travel distance (TD_{CB}^2)	−0.222***	−3.93	0.80	−19.88	**−0.106*****	**−3.10**	**0.90**	**−10.08**
Differential travel expense between CB and taxi ($TE_{\text{taxi-CB}}$)	−0.152*	−1.94	0.86	−14.08	**0.077***	**1.73**	**1.08**	**8.01**
GDP per capita at origin	**−0.742****	**−2.12**	**0.48**	**−52.40**	**0.111**	**0.29**	**1.12**	**11.73**

(continued)

Table 6.4 (continued)

	Model 1 (Inactive passengers)				Model 2 (Active passengers)			
	Coefficient	z	Odds ratio	Percentage change in odds (%)	Coefficient	z	Odds ratio	Percentage change in odds (%)
Ratio of residential–area POIs around the origin cell	**0.242**	**1.13**	**1.27**	**27.34**	**−0.602****	**−2.54**	**0.55**	**−45.22**
Population density at origin	**0.144****	**2.26**	**1.16**	**15.54**	**−0.110**	**−1.52**	**0.90**	**−10.44**
Female	**0.084**	**0.93**	**1.09**	**8.74**	−0.015	−0.14	0.99	−1.47
First subscription time	0.097***	5.85	1.10	10.14	0.020	1.38	1.02	2.00
Order of boarding station	0.053***	2.92	1.05	5.47	**0.026***	**1.71**	**1.03**	**2.59**
First subscription in winter	**0.192****	**1.98**	**1.21**	**21.11**	**0.126**	**1.27**	**1.13**	**13.39**
Constant	−4.011***	−11.61			−0.256	−0.84		
Log–likelihood	−2427				−2141			
Number of observations	3081				2116			

Note '***' indicates $p < 0.01$; '**' indicates $p < 0.05$; '*' indicates $p < 0.1$

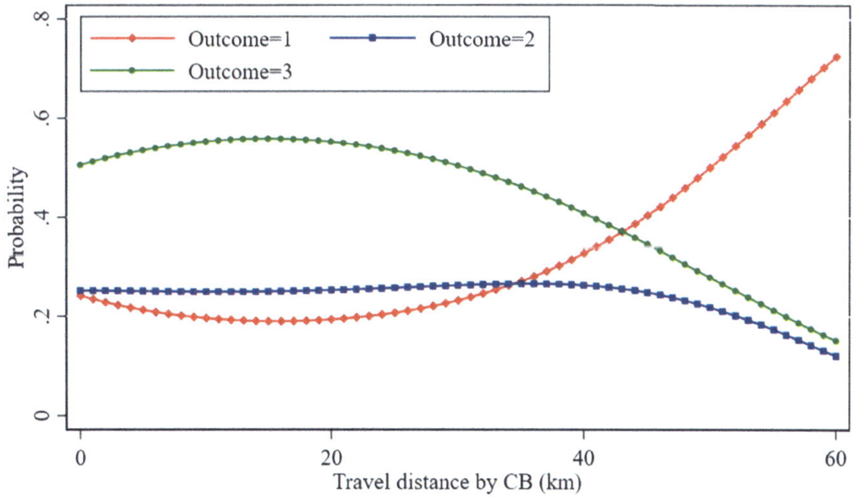

Fig. 6.3 Relationship between travel distance and probability of each category for active users (female = 1, first subscription in winter = 1, and others at means)

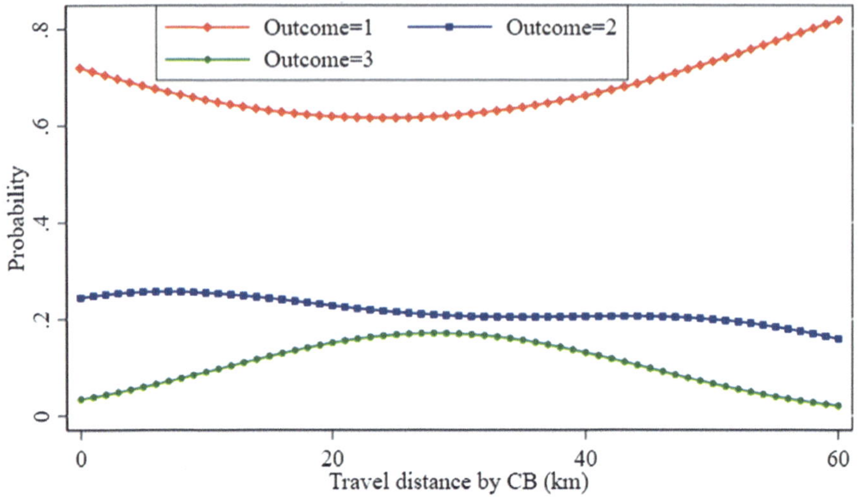

Fig. 6.4 Relationship between travel distance and probability of each category for inactive users (female = 1, first subscription in winter = 1, and others at means)

to use CB services between active users and inactive customers. For the active user group, increased GDP and decreased population density and ratio of residential POIs are associated with increased frequency of use (propensity). This effect of population density in the active user group is consistent with the finding drawn from previous studies (Wang et al. 2015, 2014) that the population density has a negative effect on the use of DRT services.

The effect of gender on the frequency of use was also completely contrary between the two groups. In particular, female users are 8.74% more likely to subscribe to the CB services frequently than male users in the inactive user group. (Wang et al. 2015) also found that female users are more likely to subscribe to the DRT services (in Lincolnshire, England) frequently when they are below pension age, while the evidence of whether the results hold for both active and inactive users is not given. In the CB case of the current study, the negative coefficient estimates in the active user group indicate that males are more likely to have a higher propensity to use CB services.

The estimates of the first subscription time consistently demonstrate the significance of the first experience in CB service, i.e., the later a user starts subscribing to the services more likely he/she is to subscribe to the services frequently. It indicates a larger effect on the frequency of use in the inactive user group than in the active user group. The positive effects reveal that service quality, such as the convenience and efficiency of subscribing/returning tickets and information availability, has been improving in the progression of the CB system.

The order of boarding stations shows similar and positive effects on the frequency of use in both active and inactive user groups. Unlike traditional public transit services, getting on earlier does not mean that the passenger has more chance of getting a seat since the CB services guarantee a seat for each passenger. On the contrary, the passengers who board the CB at later stations on the route are less likely to experience detours caused by picking up other passengers of the same CB and therefore are more likely to have a higher propensity to use CB services.

For both the active user group and inactive user group, the first-time winter subscribers are 13.39–21.11% more likely to have a higher propensity to use the CB services. It may be ascribed to the fact that the shorter waiting time and zero transfer make the CB transit services in the cold and windy winter more attractive.

6.6 Conclusions

Ascertaining the differences in subscribing behavior between active users and inactive users and developing targeted and differentiated marketing strategies accordingly will be very instrumental in increasing the passenger retention rate in DRT systems. This paper distinguished the active users and inactive users and investigated their differences in subscribing behavior through a comparative analysis, based on long-term continuous observation of a successful CB system and additional open-access geographic data sources.

Previous studies indicated a positive effect of travel distance on the use of DRT service (Kirby and Bhatt 1975; Zhang et al. 2017) but failed to provide a point of threshold in the effects of distance effects. Both two models built in this study indicate an inverted "U" shape effect of travel distance. Even though the CB services are more attractive for long-distance travel, the excessively long travel distance (longer than 28 km) is not conducive to user retention. For every RMB 10 cheaper CB trip in contrast to the taxi's, the likelihood that an active passenger subscribes to the service frequently will increase by 8%. However, the price advantage will attract novelty-seeking travelers to try CB services occasionally, but cannot convert them into high-frequency users.

An increased GDP and decreased population density and the ratio of residential POIs are associated with increased frequency of use for active users, while they show completely contrary effects for inactive users. For the inactive user group, females are more likely to travel by CB frequently. However, for the active passenger group, males are found to have a higher propensity to use CB services. There must be reasons that lead to this difference in results but the current research data does not support further exploration of the causality behind it. The passengers who board the CB at later stations on their CB route are less likely to experience detours caused by picking up other passengers of the same CB and therefore are more likely to have a higher propensity to use CB services. The shorter waiting time and zero transfer make the CB transit services in the cold and windy winter more attractive.

This study has limitations that are intrinsic to the dataset. First, individual characteristics such as household car ownership and income may also induce significant heterogeneities in travel behavior but have not been allowed to be collected owing to privacy issues. Future studies may further comprehensively factor these individual social-economic characteristics into the CB subscribing behavior modeling. Further, the dataset fetched in this study does not support the causality analysis. Therefore, the obtained association analysis results are difficult to trace back to the deterministic reasons behind.

References

Abdullah, Z. 2017. Customised bus services set to go on trial in 2018 [WWW Document]. *The Straits Times*. https://www.straitstimes.com/singapore/transport/customised-bus-services-set-to-go-on-trial-in-2018 (accessed 12.26.20).

Aldas-Manzano, J., C. Ruiz-Mafe, S. Sanz-Blas, and C. Lassala-Navarré. 2011. Internet banking loyalty: Evaluating the role of trust, satisfaction, perceived risk and frequency of use. *The Service Industries Journal* 31: 1165–1190.

Allenby, G.M., R.P. Leone, and L. Jen. 1999. A dynamic model of purchase timing with application to direct marketing. *Journal of American Statistical Association* 94: 365–374. https://doi.org/10.1080/01621459.1999.10474127.

Balcombe, R., R. Mackett, N. Paulley, J. Preston, J. Shires, H. Titheridge, M. Wardman, and P. White. 2004. *The demand for public transport: A practical guide*. Workingham, United Kingdom: TRL limited.

Bass, P., P. Donoso, and M. Munizaga. 2011. A model to assess public transport demand stability. *Transportation Research, Part a: Policy and Practice* 45: 755–764. https://doi.org/10.1016/j.tra.2011.06.003.

Brant, R. 1990. Assessing proportionality in the proportional odds model for ordinal logistic regression. *Biometrics* 46: 1171–1178.

Cao, X. (Jason). 2010. Exploring causal effects of neighborhood type on walking behavior using stratification on the propensity score. *Environment Planning A* 42: 487–504.https://doi.org/10.1068/a4269.

Cao, X. (Jason), S.L. Handy, and P.L. Mokhtarian. 2006. The influences of the built environment and residential self-selection on pedestrian behavior: Evidence from Austin, TX. Transportation (Amst) 33: 1–20.https://doi.org/10.1007/s11116-005-7027-2.

Cao, X., P.L. Mokhtarian, and S.L. Handy. 2007. Do changes in neighborhood characteristics lead to changes in travel behavior? A structural equations modeling approach. *Transportation (Amst)* 34: 535–556. https://doi.org/10.1007/s11116-007-9132-x.

Chi, C.G.Q. 2012. An examination of destination loyalty: Differences between first-time and repeat visitors. *Journal of Hospitality and Tourism Research* 36: 3–24. https://doi.org/10.1177/1096348010382235.

Chu, F. 2015. Customized bus: Cherish what you have at the moment. *Transport Business China* 8: 52–55.

Currie, G., and N. Fournier. 2020. Why most DRT/Micro-Transits fail—What the survivors tell us about progress. *Research in Transportation Economics* 83: 100895. https://doi.org/10.1016/j.retrec.2020.100895.

Dalian Municipal Bureau of Statistics. 2017. Dalian Statistic Yearbook 2017 [WWW Document]. http://www.stats.dl.gov.cn/index.php?m=content&c=index&a=lists&catid=57 (accessed 3.31.18).

Davison, L., M. Enoch, T. Ryley, M. Quddus, and C. Wang. 2012. Identifying potential market niches for Demand Responsive Transport. *Research in Transportation Business and Management* 3: 50–61. https://doi.org/10.1016/j.rtbm.2012.04.007.

Davison, L., M. Enoch, T. Ryley, M. Quddus, and C. Wang. 2014. A survey of demand responsive transport in great Britain. *Transportation Policy (Oxford)* 31: 47–54. https://doi.org/10.1016/j.tranpol.2013.11.004.

Dell'Olio, L., A. Ibeas, and P. Cecin. 2011. The quality of service desired by public transport users. *Transportation Policy (Oxford)* 18: 217–227. https://doi.org/10.1016/j.tranpol.2010.08.005.

Drengner, J., S. Jahn, and H. Gaus. 2010. Events and loyalty formation: The role of satisfaction, felt community, emotional experience, and frequency of use. In *Stand Und Perspektiven Der Eventforschung*, pp. 151–165. Springer.

Eluru, N., and S. Yasmin. 2015. A note on generalized ordered outcome models. *Analytic Methods in Accident Research* 8: 1–6. https://doi.org/10.1016/j.amar.2015.04.002.

Fu, J., D. Jiang, and Y. Huang. 2014. Gridded Population dataset of China (1km). *Change Research Data Publishing and Repository*. https://doi.org/10.3974/geodb.2014.01.06.v1.

García Gómez, B., A. Gutiérrez Arranz, and J. Gutiérrez Cillán. 2006. The role of loyalty programs in behavioral and affective loyalty. *Journal of Consumer Marketing* 23: 387–396.

Giuliano, G., and D. Narayan. 2003. Another look at travel patterns and urban form: The US and Great Britain. *Urban Studies* 40: 2295–2312. https://doi.org/10.1080/0042098032000123303.

Greene, W.H., and D.A. Hensher. 2010. *Modeling ordered choices: A primer*. Cambridge University Press.

Hu, Y., and Y. Han. 2019. Identification of urban functional areas based on POI Data: A case study of the guangzhou economic and technological development zone. *Sustainability (Switzerland)* 11: 1385. https://doi.org/10.3390/su11051385.

Huang, Y., D. Jiang, and J. Fu. 2014. 1 km grid GDP dataset of China (2005, 2010). *Global Change Research Data Publishing and Repository*. https://doi.org/10.3974/geodb.2014.01.07.v1.

Jain, S., N. Ronald, R. Thompson, and S. Winter. 2017. Predicting susceptibility to use demand responsive transport using demographic and trip characteristics of the population. *Travel Behaviour Society* 6: 44–56. https://doi.org/10.1016/j.tbs.2016.06.001.

Kieu, L.M., Y. Ou, and C. Cai. 2018. Large-scale transit market segmentation with spatial-behavioural features. *Transportation Resarch, Part c: Emerging Technologies* 90: 97–113. https://doi.org/10.1016/j.trc.2018.03.003.

Kirby, R.F., and K.U. Bhatt. 1975. An analysis of subscription bus experience. *Traffic Quarterly* 29: 403–425.

Li, J., X. Chen, X. Li, and X. Guo. 2013. Evaluation of public transportation operation based on data envelopment analysis. In *Procedia—Social and Behavioral Sciences*, pp. 148–155. Elsevier B.V. https://doi.org/10.1016/j.sbspro.2013.08.020

van Lierop, D., M.G. Badami, and A.M. El-Geneidy. 2018. What influences satisfaction and loyalty in public transport? A review of the literature. *Transport Reviews* 38: 52–72. https://doi.org/10.1080/01441647.2017.1298683.

Liu, T., and A. Ceder. 2015. Analysis of a new public-transport-service concept: Customized bus in China. *Transport Policy (Oxford)* 39: 63–76. https://doi.org/10.1016/j.tranpol.2015.02.004.

Liu, T., A.A. Cedera, R. Bologna, and B. Cabantous. 2016. Commuting by customized bus: A comparative analysis with private car and conventional public transport in two cities. *Journal of Public Transportation* 19: 55–74. https://doi.org/10.5038/2375-0901.19.2.4.

Long, J.S., and J. Freese. 2014. *Regression models for categorical dependent variables using stata*, 3rd ed. College Station, TX: Stata Press.

Le Loo, L.Y., J. Corcoran, D. Mateo-Babiano, and R. Zahnow. 2015. Transport mode choice in South East Asia: Investigating the relationship between transport users' perception and travel behaviour in Johor Bahru, Malaysia. *Journal of Transport Geography* 46: 99–111. https://doi.org/10.1016/j.jtrangeo.2015.06.011.

Ma, X. 2015. Why passengers do not accompany with customized bus? *Decision-Making* 10: 82–83.

Mageean, J., and J.D. Nelson. 2003. The evaluation of demand responsive transport services in Europe. *Journal of Transport Geography* 11: 255–270. https://doi.org/10.1016/S0966-6923(03)00026-7.

Ministry of Housing and Urban-Rural Development of the People's Republic of China (MOHURD). 2012. *Code for classification of urban land use and planning standards of development land*.

Mulley, C., and J.D. Nelson. 2009. Flexible transport services: A new market opportunity for public transport. *Research in Transportation Economics* 25: 39–45. https://doi.org/10.1016/j.retrec.2009.08.008.

Murad, H., A. Fleischman, S. Sadetzki, O. Geyer, and L.S. Freedman. 2003. Small samples and ordered logistic regression: Does it help to collapse categories of outcome? *American Statistician* 57: 155–160. https://doi.org/10.1198/0003130031892.

National Bureau of Statistics of the People's Republic of China. 2011. Tabulation on the 2010 Population Census of the People's Republic of China [WWW Document]. http://www.stats.gov.cn/tjsj/pcsj/rkpc/6rp/indexch.htm (accessed 3.31.18).

Nelson, J.D., and T. Phonphitakchai. 2012. An evaluation of the user characteristics of an open access DRT service. *Research in Transportation Economics* 34: 54–65. https://doi.org/10.1016/j.retrec.2011.12.008.

Nourbakhsh, S.M., and Y. Ouyang. 2012. A structured flexible transit system for low demand areas. *Transportation Research Part b: Methodological* 46: 204–216. https://doi.org/10.1016/j.trb.2011.07.014.

Quadrifoglio, L., R.W. Hall, and M.M. Dessouky. 2006. Performance and design of mobility allowance shuttle transit services: Bounds on the maximum longitudinal velocity. *Transportation Science* 40: 351–363. https://doi.org/10.1287/trsc.1050.0137.

Redman, L., M. Friman, T. Gärling, and T. Hartig. 2013. Quality attributes of public transport that attract car users: A research review. *Transport Policy (Oxford)* 25: 119–127. https://doi.org/10.1016/j.tranpol.2012.11.005.

Scott, R.A. 2010. *Demand responsive passenger transport in low-demand situations*. New Zealand.

Tao, T., J. Wang, and X. Cao. 2020. Exploring the non-linear associations between spatial attributes and walking distance to transit. *Journal of Transport Geography* 82: 102560. https://doi.org/10.1016/j.jtrangeo.2019.102560.

Tranberg, H., and F. Hansen. 1986. Patterns of brand loyalty: Their determinants and their role for leading brands. *European Journal of Marketing* 20: 81–109.

Transit Cooperative Research Program. 2004. Demand Responsive/ADA Traveller Response to Transportation System Changes, Synthesis Report 95. Washington, DC.

Wang, C., M. Quddus, M. Enoch, T. Ryley, and L. Davison. 2014. Multilevel modelling of Demand Responsive Transport (DRT) trips in Greater Manchester based on area-wide socio-economic data. *Transportation (Amst)* 41: 589–610. https://doi.org/10.1007/s11116-013-9506-1.

Wang, C., M. Quddus, M. Enoch, T. Ryley, and L. Davison. 2015. Exploring the propensity to travel by demand responsive transport in the rural area of Lincolnshire in England. *Case Studies on Transport Policy* 3: 129–136. https://doi.org/10.1016/j.cstp.2014.12.006.

Wang, J., T. Yamamoto, and K. Liu. 2019. Role of customized bus services in the transportation system: Insight from actual performance. *Journal of Advanced Transportation* 2019: 6171532. https://doi.org/10.1155/2019/6171532.

Wang, J., T. Yamamoto, and K. Liu. 2020a. Key determinants and heterogeneous frailties in passenger loyalty toward customized buses: An empirical investigation of the subscription termination hazard of users. *Transportation Research Part C: Emerging Technologies* 115. https://doi.org/10.1016/j.trc.2020.102636.

Wang, J., T. Yamamoto, and K. Liu. 2021. Spatial dependence and spillover effects in customized bus demand: Empirical evidence using spatial dynamic panel models. *Transport Policy (Oxford)* 105: 166–180. https://doi.org/10.1016/j.tranpol.2021.03.004.

Webb, V. 2010. *Customer loyalty in the public transportation context*. Massachusetts institute of Technology.

Weinberger, R. 2012. Death by a thousand curb-cuts: Evidence on the effect of minimum parking requirements on the choice to drive. *Transport Policy (Oxford)* 20: 93–102.

Williams, R. 2006. Generalized ordered logit/partial proportional odds models for ordinal dependent variables. *Stata Journal* 6: 58–82. https://doi.org/10.1177/1536867x0600600104.

Williams, R. 2016. Understanding and interpreting generalized ordered logit models. *Journal of Mathematical Sociology* 40: 7–20. https://doi.org/10.1080/0022250X.2015.1112384.

Yamamoto, T., J. Hashiji, and V.N. Shankar. 2008. Underreporting in traffic accident data, bias in parameters and the structure of injury severity models. *Accident Analysis and Prevention* 40: 1320–1329. https://doi.org/10.1016/j.aap.2007.10.016.

Yan, N. 2014. A customized bus market survey record [WWW Document]. http://www.chinabuses.com/buses/2014/0729/article_14514.html (accessed 3.31.18).

Yang, S., Y. Fan, W. Deng, and L. Cheng. 2019. Do built environment effects on travel behavior differ between household members? A case study of Nanjing, China. *Transprt Policy (Oxford)* 81: 360–370. https://doi.org/10.1016/j.tranpol.2017.12.006.

Yao, Y., X. Li, X. Liu, P. Liu, Z. Liang, J. Zhang, and K. Mai. 2017. Sensing spatial distribution of urban land use by integrating points-of-interest and Google Word2Vec model. *International Journal of Geographical Information Science* 31: 825–848. https://doi.org/10.1080/13658816.2016.1244608.

Zhang, J., D.Z.W. Wang, and M. Meng. 2017. Analyzing customized bus service on a multimodal travel corridor: An analytical modeling approach. *Journal of Transportation Engineering, Part a: Systems* 143: 04017057. https://doi.org/10.1061/jtepbs.0000087.

Chapter 7
Key Determinants and Heterogeneous Frailties in Passenger Loyalty

Abstract Long-term passenger subscription is vital to the survival and operation of the customized bus (CB) system, which is a demand-driven and user-oriented transit service. A better understanding of passenger loyalty toward the CB service will help provide better operation. The urgent and outstanding issue is how to incorporate the unobserved heterogeneity in loyalty—in other words, how to reflect the effects of the frailty to terminate subscription. This study fills the research gap through an empirical study in Dalian, China. Three different survival models are developed to investigate the mechanism of subscription behaviors, among which the shared frailty model considering the unobserved heterogeneity is demonstrated to be the most appropriate. The results indicate that the historical purchase characteristics are the most important to CB user loyalty modeling and forecasting. Males are more sensitive than females to the number of intermediate stations because of the potentially increased uncertainty in waiting time related to the intermediate stations. The heterogenous frailties resulting from the heterogeneity of the perceptible service quality in terms of convenience and efficiency in subscribing/returning tickets and information availability in the progress of the CB system significantly contribute to user loyalty deviations.

Keywords Demand responsive bus · Subscription behavior · User loyalty · Demand evolution · Shared frailty model

7.1 Introduction

An emerging demand-driven, user-oriented transit service, known as a customized bus (CB) service, holds the promise of alleviating urban traffic congestion, especially under the extremely tight road capacity constraints during peak hours. In contrast to conventional public transportation (PT) services, CBs aggregate users with similar origins, destinations, and departure/arrival times (Liu and Ceder 2015a). The aim of a CB service is to provide point-to-point transit services (with one person per seat) without any transfers between the appointed boarding and deboarding stations. Thus,

J. Wang et al., *Customized Bus: User Travel Behavior and Demand Evolution*, Urban Sustainability, https://doi.org/10.1007/978-981-96-3488-0_7

user requests are assigned as a top priority. The CB service has gained popularity in many large and medium-sized cities in China since it was first launched in Qingdao (August 2013), similar to other on-demand ride services (Ke et al. 2017).

A profound understanding of the subscription behavior is vital not only for service improvement but also for maximizing user retention. Due to the high operation costs, CB routes with high seat occupancy rates, for example, greater than 80%, have been launched. CB routes with relatively low seat occupancy rates are partially adjusted, combined, or may be canceled if the occupancy rate in operation is less than 50% for several months. The survival of a CB route heavily depends on the stability of the subscription quantity and the duration for each user, which represent the level of user loyalty. In reality, even though the service providers have made concessions on the occupancy rate and developed strategies such as combining similar routes to retain passengers, there have been drastic decreases in the number of CB service users after its prevalence in many cities (Yan 2014); (Chu 2015); (Ma 2015), such as Guangzhou, Hangzhou, Nanjing, Xiamen, and Harbin. Many launched routes have been canceled, and many users have unsubscribed from the service and returned to their previous modes of transportation, leading to a substantial waste of resources and the disappointment of many loyal patrons. The drastic decrease in the number of CB users has been mainly attributed to the following reasons: (1) passengers gradually stopped using the service after seeking novelty with regard to this new concept of personalized transit service or just finishing the follow-the-leader behavior; (2) the absence of demand forecasting techniques and flexible route designing methods that could satisfy the rapid increase in new users, while the launched routes could not attract enough passengers to maintain operation without discount coupons; and (3) the prereservation of the CB service and compliance with the predetermined schedule, which is not as flexible as using a conventional high-frequency PT system.

Essentially, these decreases show the vicious transformation of users' loyalty toward the service that they have subscribed to. Ensuring the survival of the launched routes is the only win–win solution for both operators and loyal users. Feedback from the viewpoint of user subscription behavior is a good indicator of user satisfaction toward the supplied service and will provide practical references for policy-making and market orientation. Factors leading to the loss of loyalty can be targeted with intensive strategies to maximize user retention. Therefore, ascertaining the key determinants of passenger loyalty, as well as modeling and forecasting the subscription duration, are urgent and critical events in the development of the CB system.

There is a paucity of studies that focus on investigating passenger loyalty directly by using their actual purchase records (Trépanier et al. 2012). Additionally, there are even fewer studies on user loyalty toward the CB transit service. This work aims to fill this gap by investigating CB user loyalty through a survival analysis based on their long-term subscription behaviors. The survival models, or so-called duration models, are employed to model user loyalty with factors drawn from historical subscription behaviors, travel attributes, and individual characteristics. The use of the survival model enables the analysis of factors that mainly drive users' loyalty and facilitates the investigation of the mechanism of subscription behaviors. From the viewpoint of system operation, analyzing passenger survival time is an important benefit because it can help measure user loyalty at different stages and respond to strategy changes for further improvement of the service.

The rationale of the CB service is illustrated in Fig. 7.1. The CB aggregates users with similar origins, destinations, and departure/arrival times, which can be clustered by demarcating the demand distribution into OD bundles. Each OD bundle is composed of an origin zone, a destination zone, and a passing-by corridor. The ladder-shaped origin/destination demarcations are beneficial for shortening the detour distance by picking up and dropping off passengers. The boarding and alighting stations are designed to be located close to the users' residences and workplaces. Moreover, there are no stations located in the span of the passing-by corridor, thereby making the route flexible and ensuring that the drivers follow the vehicle navigation system to minimize the travel time by following the shortest-time path. Users can subscribe to CB services by using easy-to-access websites and smartphone applications. The CB service has rapidly become popular in China due to its simplicity, convenience, comfort, and customized features.

The remainder of this study is organized as follows. Section 7.2 reviews key studies relevant to the investigation and modeling of passenger loyalty in transit systems. Detailed descriptions of the study area and the data collection of the CB system are given in Sect. 7.3. The methodology employed in this study and the models built are described in detail in Sect. 7.4. Finally, the results and discussions, as well as the conclusions drawn based on the key findings of this study, are presented in Sect. 7.5 and Sect. 7.6, respectively.

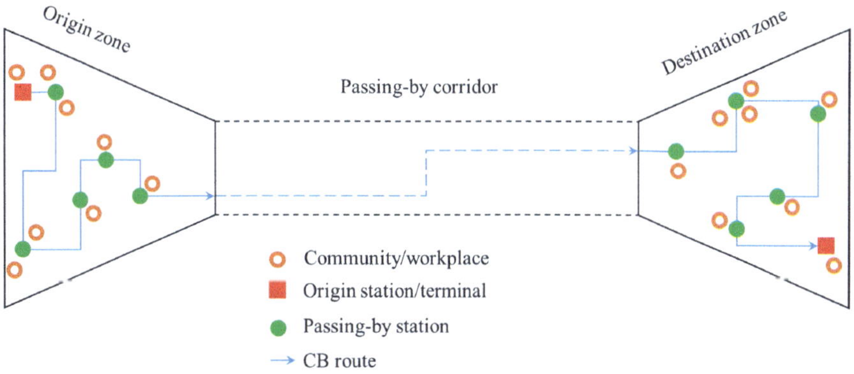

Fig. 7.1 Typical example of a CB route

7.2 Literature Review

Customer loyalty toward a specific product or service is defined as the repetitive purchase behavior that reflects the decision to continue buying the same product or service (Jacoby and Chestnut 1978). Customer loyalty can be classified into behavioral loyalty and attitudinal loyalty (Webb 2010); (Odin et al. 2001). Behavioral loyalty is a person's repetitive selection of a certain brand over the competition (Odin et al. 2001), whereas attitudinal loyalty is associated with the psychological or emotional commitment toward a specific product, ignoring the actual choice of an individual.

Ridership retention in CB systems is still a challenge for almost all operators. Passenger loyalty was always considered to be relevant to passenger satisfaction with these services. There are studies that address user loyalty in PT systems from the standpoint of user satisfaction (or at least taken as a factor). For example, (Jomnonkwao et al. 2015) pointed out that user satisfaction and perceived value of bus services directly and significantly affect loyalty levels of patrons. (Wen et al. 2005) investigated passenger loyalty toward intercity bus services and identified important factors influencing loyalty through a structural equation model (SEM). They found that satisfaction has the most significant influence on passenger loyalty. (Lai and Chen 2011b), also by using SEM models, examined the interactions between satisfaction and behavioral intentions, and concluded that passenger behavioral intentions or loyalty significantly rely on passenger satisfaction. Furthermore, service reliability/punctuality, safety, user perceptions, and service quality are also predictors of passenger loyalty in PT systems (Bass et al. 2011; Figler et al. 2011; Imaz et al. 2014; Redman et al. 2013). In practice, the growing difficulties in maintaining highly satisfactory transit services while simultaneously improving the mobility of citizens are encountered. Fortunately, some researchers have dedicated themselves to the determination of solutions that maximize user satisfaction and optimize operating schemes, such as the ontology-based framework to support performance monitoring (Benvenuti et al. 2017), joint optimization model of passenger-to-vehicle assignment (Tong et al. 2017a), and computerized routing and scheduling (Marković et al. 2015).

Most of the aforementioned studies investigated the passenger intention and anticipated user loyalty toward changes in policies based on data collected from questionnaire surveys and face-to-face interviews. These data provide a method to investigate the decision-making psychology of respondents under different scenarios with two or more alternatives, where the differences between every two alternatives are emphasized. Nevertheless, the information collected in the stated preference surveys reflects a probabilistic choice made by a respondent based on the given attributes for each alternative but not his/her actual purchase behavior. Although the reviewed preference surveys can catch the actual behaviors of respondents, one can hardly follow all the respondents for a long period to capture their continuous choice behaviors due to the substantial cost.

Previous studies barely investigated passenger loyalty directly using their reuse records, despite the availability of a significant amount of data collected from smart

PT cards. Concerns such as privacy, record information errors, and absence of alighting location information could be responsible for the limited implementation (Ghasemzadeh et al. 2014)(Trépanier et al. 2012). Nevertheless, some researchers have emphasized that forecasting whether a user will be active in the future depends on his/her past purchase behavior (Allenby et al. 1999). A lack of historical purchase records has been taken as a limitation when investigating user loyalty (Guillén et al. 2012). On the basis of these repetitive purchase data, the survival duration of customers has been widely used as a measurement of customer loyalty, and the survival model is recommended as a well-adapted and more effective method in previous literature (Guillén et al. 2012) (DuWors and Haines 1990);(Jin and Guo 2018). Fortunately, this work successfully collected the repetitive subscription records of each user for more than two years. Thus, survival analysis is a natural choice for modeling the censored data (Kleinbaum and Klein 2010). Because we cannot follow up with all the users forever, sometimes the time of subscription termination is not observed within a finite study period and therefore is subjected to right censoring. In practice, the survival models have been widely used in the transportation field. Some typical studies include investigation on the delay of commuters' work-to-home trip to avoid traffic congestion (Mannering and Hamed 1990), accident duration prediction (Chung 2010), and the length of time spent on activities such as shopping and recreational/social visits studied using the Cox model and accelerated failure time models (Kharoufeh and Goulias 2002) (Kim et al. 1993).

In this study, the "survival time" of a user on the CB service is the total length of time that the user subscribes to the CB service repeatedly. The risk of terminating the CB subscription over a period naturally indicates the level of user behavioral loyalty. A higher risk of terminating the subscription over a period represents a lower level of loyalty at the given time for the individual, and vice versa.

7.3 Research Area and Data Collection

The study area chosen for this work is Dalian, a coastal city in Northeast China. There are different PT modes available in Dalian, such as subways, light rail transit, and regular buses. The rail transit system in Dalian is not well developed. There are two subway lines that primarily connect the functional areas of the city center, whereas the other two light railway lines connect emerging city outskirt centers to the main city areas.

In March 2014, the first CB route was launched in several travel corridors in Dalian city, initiated by the state-owned PT company. After that, the CB services spread to more areas thanks to the better level of service (Wang et al. 2019a) and the employment of environmental-friendly electrical buses (Liu et al. 2017, 2018). More CB routes were subsequently made available for reservation, and a few private companies collaborated with the government to provide more CB routes for citizens. Unlike the conventional PT services, CB is a type of subscription transit service where the customer gets access to the service by subscribing to the preferred route using a smartphone application or an easy-to-access website.

In addition, there are two types of subscriptions: daily subscriptions and monthly subscriptions. A daily subscription is a "ticket" for a one-way trip or round trip in a day, whereas the monthly subscription is valid for a specific month, including one-way and round trips. The daily subscriptions mainly provide service for random demands alongside some highly demanded corridors, whereas the monthly subscription is mainly the choice of regular CB users. In the Dalian CB system, monthly subscribers rapidly dominated the passenger groups after the service was made officially available (Wang et al. 2019a). The subscription behavior of monthly ticket users has a greater impact on the demand evolution of the CB system and therefore is the focus of this work. Subscription records and related route information from August 2015 to September 2017 of a CB system were acquired from the enterprise, which played a vital role in initiating and operating the CB services in Dalian.

Upon data scrubbing, 6912 registered users with monthly CB subscriptions were found, without any loss of information. The data includes subscription records for all of these users as well as corresponding coordinates of boarding/deboarding stations of their subscribed CB routes. The route information was collected by recording the coordinates of all stations in order as well as the designated departure time for all CB routes.

Figures 7.2 and 7.3 illustrate the changes in the number of passengers in the research period and the observed survival rate by the time of the subscription renewal, respectively. In particular, the observed survival rate is measured by the percentage of passengers (originally generated at the same time) who resubscribe to the service at each subscription renewal. Essentially, this rate decreases over time. The monthly subscription became officially available in August 2015 for the first time. In December 2015, the number of registered passengers first reached more than 2500 and then remained relatively steady. The largest number of users began subscribing to the CB service in December 2015 and contributed the most in terms of the number of users retained, mainly driven by user loyalty. However, there is no evidence showing that any aggressive marketing strategy resulted in the spike in new users. Subsequently, the retained users turned out to be the main component of passengers, and only a small number of new users were generated.

After the trial operation period, which provided a large number of discount coupons or lucky free tickets (the operator randomly distributed free tickets of one-day CB trips to a limited number of users for experience marketing) from April 2014 to March 2015, the Dalian CB operator reduced the number of coupons significantly. To make the launched routes more vital and sustainable, and thus maximize passenger retention, the operator designed the policy for route launching and canceling. This policy helped to recognize the best routes in operation with enough travel demand, while in the experiment, the operator endeavored to make the policy more flexible. The CB routes with relatively lower seat occupancy rates were partially adjusted, combined, or canceled if the occupancy rate in the operation was less than 50% for several months. Therefore, with new users (as shown in Fig. 7.2), the cases of route cancellation several months after the launch were quite rare. Thus, the primary

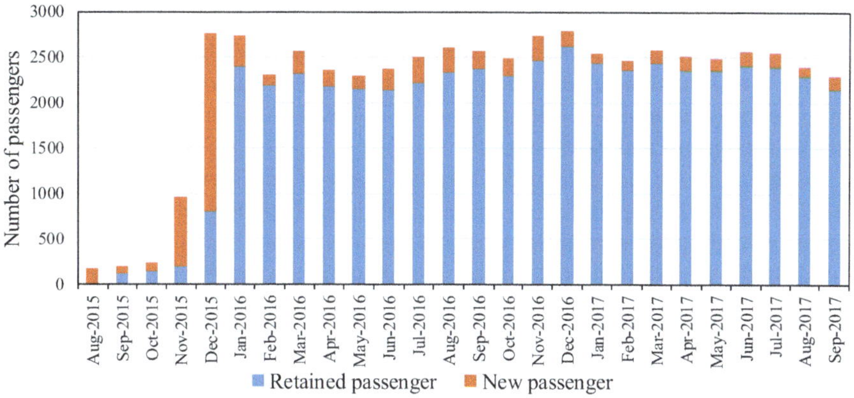

Fig. 7.2 Changes in the number of passengers during the research period

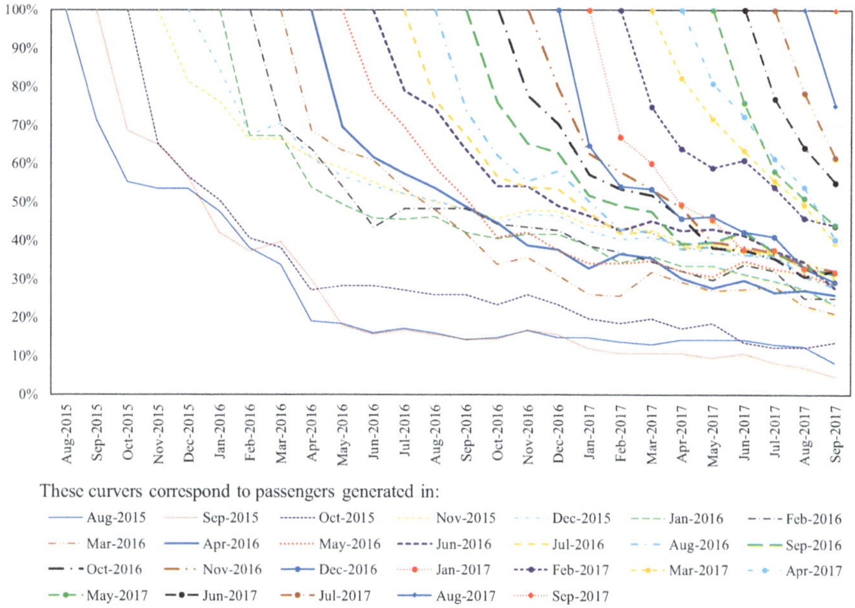

Fig. 7.3 Observed survival rate by the time of subscription renewal

reason for terminated subscriptions is the loss of user loyalty. To maximize user retention, it is critical to identify the key determinants of user loyalty and ascertain the mechanism of their subscription behavior.

However, detailed information on the actual timetable at each station of the CB routes is unavailable. To obtain the actual travel attributes of the CB trips as well as other available modes, all routes for each OD pair were replanned for CBs, PT, and taxis, and the corresponding data were collected. These works were accomplished by coding on the Baidu application programming interface (API), which is a service provided by Baidu Inc., and enables developers to interact with the Baidu map servers directly. It is a functional and widely used instrument to acquire a large amount of real-time travel information, such as travel time, travel speed, and traffic conditions, with known origins and destinations.

The program for the replanned CB routes was run on Thursday, November 30, 2017, to obtain the departure and arrival times at each station (also used for other modes) according to the planned departure times at the origin stations for each CB. Based on the departure time acquired at each station, a second program was run on Thursday, December 14, 2017, to obtain the travel attributes of other modes. In particular, the travel information was derived at the same time of the day and on the same day of the week to minimize variations in the traffic conditions. Finally, the travel time (*TT*), travel distance (*TD*), average travel speed (*ATS*), and travel expense (*TE*) were collected and calculated for each OD pair in all three modes. For more details about the settings in route replanning programs, please refer to our previous study (Wang et al. 2019a).

In particular, the travel expense of PT and taxi were obtained using the program on the Baidu map API, based on the charging system of PT and taxi in Dalian, respectively. The expense for passengers traveling by CB was determined based on the distance tariff scheme (Liu and Ceder 2015a; Wang et al. 2019a).

7.4 Methodology

7.4.1 Loyalty of CB Users

A survival analysis uses a collection of statistical procedures for data analysis, for which the outcome variable of interest is the time until an event occurs (Kleinbaum and Klein 2010). It is popularly used in many areas such as medicine, biology, engineering, and economics. In this study, the event is the termination of the CB service subscription, which indicates the loss of loyalty from a specific passenger. When a user has stopped subscribing for a long time, usually longer than his/her historically maximal interval time, it can be deemed that this user has terminated the subscription (no longer subscribes to the service) and will not be loyal anymore (the event occurs).

The survival function $S(t)$ is fundamental to the survival analysis, given that the probability that a user stays active in subscribing to the CB service, assumed as T, is longer than some specified time t.$f(t)$ and $F(t)$ are assumed as the probability density function and cumulative distribution function of the continuous random variable T, respectively. Thus, the probability that a registered user stays active (survival time T) is either longer than or equal to t is given by (Washington et al. 2003):

$$S(t) = Pr(T \geq t) = 1 - F(t) \tag{7.1}$$

Another fundamental formula is the hazard function $h(t)$, which provides the instantaneous potential per unit time for the termination of subscription (event) to occur, given that the user has stayed active up to time t. The hazard function is sometimes called a conditional failure rate (Washington et al. 2003), given by:

$$h(t) = \frac{f(t)}{S(t)} = \lim_{\Delta t \to 0} \frac{\Pr(T < t + \Delta t | T \geq t)}{\Delta t} \tag{7.2}$$

Each of the two functions can be derived from the other because of their clearly defined relationship. The hazard function can be obtained from the survival function; for instance,

$$h(t) = -\left[\frac{dS(t)/dt}{S(t)}\right] \tag{7.3}$$

In this study, user loyalty was measured by modeling the hazard function related to the probability of terminating subscription by an individual at time t. Lower risk of subscription termination indicated a more loyal CB service user.

7.4.2 Cox Proportional Hazard Model

There are several models available for survival analysis. These models include the Kaplan–Meier (K–M) method, the Cox proportional hazards model, and the accelerated failure time model. The K–M model is a non-parametric method in which there is no chance of taking other associated covariates into consideration. The Cox proportional hazards model is a semiparametric method in which a part of hazard function can be explained by covariates, while the distribution of survival time is unspecified. The accelerated failure time model, which is the representative of parametric methods, considers the effects of covariates and distribution of survival time, simultaneously (Lee and Wang 2003). In practice, it may not be completely certain if a given parametric model is appropriate (Kleinbaum and Klein 2010). Thus, the semiparametric setting of the model and robust estimates of coefficients make the Cox proportional hazards model popular for a wide variety of data situations.

The Cox proportional hazards model assesses the effect of risk factors on the survival time and works for both quantitative predictor variables and categorical variables. This model provides a method for examining the effect of specified factors on the rate of occurrence of an event (termination of subscription) at a specific time. The Cox model is expressed by the hazard function $h(t)$, indicating the risk of the event happening at time t.

$$h_i(t|\boldsymbol{X}_i) = h_0(t)exp\left(\boldsymbol{X}_i^T \boldsymbol{\beta}\right) \tag{7.4}$$

where $h_i(t|\boldsymbol{X}_i)$ is the hazard at time t for individual i with a given specification of a set of covariates denoted by $\boldsymbol{X}_i$. The first part $h_0(t)$ is the unspecified baseline hazard function, which involves t but does not contain any covariates, thus corresponding to the value of the hazard when all variables are equal to zero and indicating that the risk varies over time. The second part is the exponential expression e to the joint linear effect of a series of covariates.

For the interpretation of β_k (the coefficient for k-th explanatory variable), the hazard ratio is formulated as:

$$R_{x_k} = e^{\beta_k \Delta x_k} \tag{7.5}$$

where R_{x_k} is the hazard ratio of variable x_k, adjusted for other variables, and Δx_k is the change in variable x_k. When the value of β_k is larger than zero, the hazard ratio is greater than one, indicating that a Δx_k increase in x_k will increase the hazard to meet the event by $e^{\beta_k \Delta x_k}$, thereby reducing the survival duration.

7.4.3 Shared Frailty Model

In general, correct inferences can be obtained based on Cox proportional hazard models when using independent and identically distributed samples. Despite controlling for known risk factors, individuals may be exposed to different risk levels due to the unobserved heterogeneities in different clusters. Therefore, the subjects from the same cluster may behave more cohesively than those from different clusters. The shared frailty model is an extension of the Cox proportional hazards model that provides a way to account for the clustering heterogeneities in survival data (Wienke 2011). The key idea of the shared frailty models is that individuals in different clusters have different frailties, and the frailest will die earlier than the less frail. Frailty is always taken as an unobserved random factor that multiplicatively modifies the hazard function of a cluster of individuals.

The hazard function for a shared frailty model is generally formulated as follows:

$$\begin{aligned} h_{ij}\left(t|\boldsymbol{X}_{ij}\right) &= h_0(t)u_j exp\left(\boldsymbol{X}_{ij}^T\boldsymbol{\beta}\right) \\ &= h_0(t)e^{\sum_{k=1}^{m}\beta_k x_{ijk}+v_j} \end{aligned} \tag{7.6}$$

where u_j is the frailty included for individuals in cluster j, $j \in \{1, 2, \ldots, n_j\}$. Frail users with high values of u_j will tend to terminate the service earlier than those with lower values of u_j. As the formulation shows, the frailty can be interpreted and estimated as the random effect v_j in the general regression form, and it can be assumed to have an independent and identical statistical distribution. In the study, the distribution of frailty is modeled as a Gaussian distribution with a mean of zero and variance of σ^2. The other terms are the same as those in Eq. (7.4).

It is difficult to estimate the parameters in Eqs. (7.4) and (7.6) using ordinary likelihood since the baseline hazard function $h_0(t)$ is completely unspecified (Cox 1972). Instead of the full likelihood, (Cox 1972) conceived a partial likelihood to estimate the effects of covariates without the need to consider the baseline hazard. For the particular occurrence of the event at the time t_i, , the probability (taking Eq. (7.4) for instance) that the event happens on an individual can be written as:

$$\frac{h_0(t_i)exp\left(\boldsymbol{X}_i^T\boldsymbol{\beta}\right)}{\sum_{p:tp>t_i} h_0(t_i)exp\left(\boldsymbol{X}_p^T\boldsymbol{\beta}\right)} \tag{7.7}$$

where $\{p:t_p > t_i\}$ is the set of individuals at risk at time t_i, including users whose termination time is at least t_i. The baseline hazard function can be canceled out, so the joint partial likelihood function is then formulated as:

$$L(\boldsymbol{\beta}) = \prod_{i=1}^{n}\left[\frac{exp\left(\boldsymbol{X}_i^T\boldsymbol{\beta}\right)}{\sum_{p:tp>t_i} exp\left(\boldsymbol{X}_p^T\boldsymbol{\beta}\right)}\right]^{\delta_i} \tag{7.8}$$

where the termination of subscription is indicated by $\delta_i = 1$, while the $\delta_i = 0$ denotes censoring time.

A generalized form of the goodness of fit for hazard models was originally proposed by (Cox and Snell 1989). Nevertheless, (O'Quigley et al. 2005) found that Cox's R^2 decreases dramatically with an increase in the percentage of censoring and further developed this method by replacing the number of observations n with the number of uncensored events h:

$$R^2 = 1 - exp\left[\frac{-2LL\left(\widehat{\boldsymbol{\beta}}\right)+2LL(0)}{h}\right] \tag{7.9}$$

where $LL\left(\widehat{\boldsymbol{\beta}}\right)$ is the log-likelihood of the fitted model; $LL(0)$ is the partial log-likelihood for the null model; and h is the number of users who have been observed to terminate the subscription during the research period.

Additionally, based on information theory, the Akaike information criterion (AIC) is employed to deal with the trade-off between the goodness of fit and the simplicity of the model (Akaike 1998). If a model loses less information (the AIC), it is deemed to have better performance. The AIC is given by:

$$AIC = 2q - 2LL\left(\widehat{\boldsymbol{\beta}}\right) \tag{7.10}$$

Where q is the number of parameters to be estimated in the model.

7.4.4 Variable Selection and Modeling

As discussed in the introduction section, (1) the limited perception of the new concept of personalized transit service, (2) the absence of flexible demand forecasting and route design method, and (3) the strict reservation rule and riding schedule of the CB in contrast to traditional PT are the three main reasons that led to a decrease in CB users. To identify the determinants of user loyalty and shed light on the mechanism of their effects, three groups of variables are required in terms of these issues: (1) historical subscribing behavior characteristics, (2) travel attributes, and (3) individual passenger characteristics.

First, the characteristics of historical purchase behavior are important for identifying the novelty-seeking and follow-the-leader users as well as those who are more loyal to the CB service. There are researchers who indicated that the past purchase characteristics of users are indispensable in the prediction of user retention (Allenby et al. 1999: Kim and Mannering 1997) focused on the historical behavior characteristics in their study by including proxy variables of "occurrence dependence" and "lagged duration dependence" in the survival analysis and indicated the important correlations between previous behavioral characteristics and the current duration. In particular, a user who continuously subscribed to the service for more than two years, despite several intervals in between, is strongly loyal to the CB service. In contrast, a user who experiences a significant difference between the expected and actual CB service will naturally quit this new method of transportation after a few trials. The following behavioral characteristics of subscription are extracted from our dataset: (1) consecutive subscription duration, (2) maximal interval time between two sequential subscriptions, and (3) first subscription time for each patron. Accordingly, we expect that:

H1: The longer the period of consecutive subscription for a user, the more loyal he/she is to CB services.

H2: Subscribers who have experienced a longer period of break and continued to subscribe are more loyal to CB services.

To investigate the influence of the inflexibility of monthly subscriptions (valid for all workdays in a month), led by strict reservation rules and riding schedules, the daily subscription experience was included in the survival model. In particular, we focused on the loyalty difference in groups with and without daily subscriptions utilized before a user registered for the monthly subscription. Thus, it can be proposed that:

H3: The users who have not utilized the daily subscription are more loyal to CB services.

The absence of reliable demand forecasting and a reasonable route design directly influences the service quality reflected by the attributes of trip, which is considered to be a critical factor for passenger loyalty in a PT system, and the differences in travel attributes with other modes are emphasized. Moreover, the differences in the travel attributes between CBs and other available travel modes reveal the motivations of users shifting their commuting mode to/from CBs. For instance, a person is more likely to shift from PT to CB because of the shorter travel time. Individuals are more likely to shift from taxi to CB because it costs much less. Therefore, the difference in travel attributes, including the travel expense difference between taxi and CB, the travel expense difference between CB and PT, the travel time difference between PT and CB, and the travel speed difference between CB and PT, were measured and included as covariates in the survival models. Other variables related to the travel attributes are excluded due to testing as nonsignificant. Moreover, the availability of rail transit service for a passenger's trip was investigated and included as a dummy variable. The order of the boarding station for each passenger is also included in the models.

Furthermore, individual characteristics, such as gender and age, may also have impacts on the use of the demand-responsive-transit service (Davison et al. 2012; Mageean and Nelson 2003; Wang et al. 2015). Similar to other systems (Mageean and Nelson 2003), females are the dominant users in the Dalian CB system (more than 75%). This finding may be attributed to the lower vehicle and driver license ownership in females in contrast to males (Davison et al. 2012). On the other hand, females have more concerns about safety and security during their trips (Enoch et al. 2004). Therefore, the point-to-point, one-seat guaranteed, and uncrowded CB transit services are more attractive to females who typically commute by conventional PT services. In this study, only gender information was collected for privacy issues, but we believe that there must be some other individual characteristics that contribute to the variation in user loyalty.

The subscription duration and the variables selected to investigate user loyalty are summarized in Table 7.1. Among the 6912 registered users, the event was observed in 4050 instances during the study period. In particular, 2771 observations were right censored because they had not terminated the CB subscription before the study ended. There are 91 users for whom the event was observed just after the cancelation of their routes during the study period. In these cases, the cancelation of their route may be another reason for users to terminate the subscription aside from the loss of loyalty. Thus, these cases are also regarded as censored data, and their observed subscription durations are taken as their survival durations. In this study, all censoring mechanisms were independent of the event process; therefore, it is appropriate to use a survival analysis (Washington et al. 2003).

The Cox proportional hazard model and shared frailty model are employed to model the survival duration of users. The effects of all the factors discussed above are examined in the survival models. The parameters in terms of the subscribing behavior characteristics and most factors in the travel attributes directly demonstrate their meaningful principles. However, a meaningful interpretation of the individual characteristics cannot be drawn easily. The effect of a factor may depend on the value of another variable. In this case, it may be difficult to ascertain how this variable affects the hazard ratio without including the interaction effect (Aiken et al. 1991), especially for categorical variables. The conclusion that "females are more loyal than

Table 7.1 Summary of variables and descriptions

Variable (Symbol)	Description	Mean	Std. Dev	Median	Min	Max
Duration	Total duration of subscription (in months)	9.89	7.86	7	1	26
Survival status Censored (2862) Termination (4050)	It is equal to 1 when the passenger has already terminated the subscription (event) and 0 when he/she stays active at the end of study period or the termination may be caused by other reasons such as route cancelation within the study period	–	–	–	–	–
Individual characteristics						
Gender dummy (*F*) Male (1704) Female (5208)	It is equal to 1 for female and 0 for male	–	–	–	–	–
Historical subscribing behavior characteristics						
Daily subscription utilized Utilized (498) Not utilized (6414)	It is equal to 1 when a passenger utilized daily tickets before he/she used a monthly ticket for the first time; else, it is 0	–	–	–	–	–
Consecutive subscription duration (*CT*)	The maximal duration of unceasing subscription without any break during the research period (in months)	6.95	6.11	5	1	26
Order of boarding station (*OS*)	The order of the appointed boarding station of the specified CB	3.21	2.95	2	1	16
Maximal interval time (*MT*)	The maximal interval time between two sequential subscriptions during the research period (in months)	1.2	2.49	0	0	21
First subscription time (*FT*)	The month when a user subscribes to the monthly subscription for the first time (June 2015 is taken as the baseline)	9.19	4.83	6	0	25
Route and travel-related attributes						

(continued)

Table 7.1 (continued)

Variable (Symbol)	Description	Mean	Std. Dev	Median	Min	Max
Differential travel expense between taxi and CB ($TE_{Taxi-CB}$)	$TE_{Taxi-CB} = TE_{Taxi} - TE_{CB}$(in RMB)	21.99	16.05	18.8	3.2	128.4
Differential travel expense between CB and PT (TE_{CB-PT})	$TE_{CB-PT} = TE_{CB} - TE_{PT}$(in RMB)	4.94	2.09	4.4	−0.2	15.2
Differential travel time between PT and CB (TT_{PT-CB})	$TT_{PT-CB} = TT_{PT} - TT_{CB}$(in minutes)	31.55	18.56	30.82	− 55.18	103.27
Differential travel speed between CB and PT (ATS_{CB-PT})	$ATS_{CB-PT} = ATS_{CB} - ATS_{PT}$(in km/h)	11.17	5.36	10.42	− 7.96	33.01
Rail transit service (RA) Available (3747) Unavailable (3465)	Dummy variable; it is 1 when the rail transit services are available for the passenger's commuting trip in part or the entire way based on the recommended three route plans (the best routes with 1) shortest time, 2) shortest distance, or 3)minimum thansfer) from the Baidu map; else it is 0	–	–	–	–	–

males" seems more of the phenomenon but does not reveal the principle behind this pattern. The travel attributes significantly demonstrate the different contributions to the satisfaction level between females and males due to their different requirements on the service level (Tjeendra et al. 2010). In this study, we have assumed:

H4: Females and males have different requirements regarding travel attributes.

To acquire unbiased estimates (Vatcheva et al. 2015), the moderating effects between gender and all travel attributes were examined, and the moderating effect between the female dummy and the order of the boarding stations was indicated to be statistically significant and was subsequently considered in the models.

Similarly, the availability of the rail transit services may influence user satisfaction with travel attributes such as travel speed. We hypothesize:

H5: Whether rail transit services are available during commuting trips of passengers will lead to heterogeneities in satisfaction on travel attributes.

Thus, the moderating effects between the availability of the rail transit service and these variables were examined, and the interaction term between the availability of the rail transit and the differential travel speed (between CB and PT) was found to be statistically significant. First, we examined the joint linear effects of these covariates and the corresponding moderating effects (called the Cox model for convenience) in order to demonstrate a fundamental insight into passenger loyalty:

$$\begin{aligned} h_i(t|\boldsymbol{X}_i) = h_0(t)exp(&\beta_1 F_i + \beta_2 CT_i + \beta_3 OS_i + \beta_4 MT_i + \beta_5 RA_i + \beta_6 TT_{PT-CB,i} \\ &+\beta_7 TE_{Taxi-CB,i} + \beta_8 TE_{CB-PT,i} + \beta_9 ATS_{CB-PT,i} + \beta_{10} OS_i \cdot F_i \\ &+\beta_{11} RA_i \cdot ATS_{CB-PT,i}) \end{aligned} \tag{7.11}$$

Whether a user had utilized a daily subscription before he/she first joined the monthly subscription may result in a difference in loyalty to the CB service. Daily subscription is more flexible but has a higher unit price compared to the monthly subscription. Thus, daily subscription provides a chance for prospective users to experience the service before joining as a regular user and therefore leaves an initial impression of the CB service on the users. Due to the violation of the proportional hazards assumption, a stratified model is specified with baseline hazard functions stratified by the daily subscription experience.

Daily subscription utilized before trying the monthly subscription:

$$\begin{aligned} h_{i,1}(t|\boldsymbol{X}_i) = h_{0,1}(t)exp(&\beta_1 F_i + \beta_2 CT_i + \beta_3 OS_i + \beta_4 MT_i + \beta_5 RA_i + \beta_6 TT_{PT-CB,i} \\ &+\beta_7 TE_{Taxi-CB,i} + \beta_8 TE_{CB-PT,i} + \beta_9 ATS_{CB-PT,i} + \beta_{10} OS_i \cdot F_i \\ &+\beta_{11} RA_i \cdot ATS_{CB-PT,i}) \end{aligned} \tag{7.12}$$

Daily subscription not utilized before trying the monthly subscription:

$$\begin{aligned} h_{i,2}(t|\boldsymbol{X}_i) = h_{0,2}(t)exp(&\beta_1 F_i + \beta_2 CT_i + \beta_3 OS_i + \beta_4 MT_i + \beta_5 RA_i + \beta_6 TT_{PT-CB,i} \\ &+\beta_7 TE_{Taxi-CB,i} + \beta_8 TE_{CB-PT,i} + \beta_9 ATS_{CB-PT,i} + \beta_{10} OS_i \cdot F_i \\ &+\beta_{11} RA_i \cdot ATS_{CB-PT,i}) \end{aligned} \tag{7.13}$$

Moreover, a previous study indicated that satisfaction plays a more important role in loyalty expressed by first-time users than that of repeater subscribers (Chi 2012). As an emerging service of transportation, the level of service has naturally changed as the CB system has progressed. The service operators gradually improved the service quality to attract more potential users based on operation experience. Thus, the resulting heterogeneity on satisfaction for the first time, when a user purchases the monthly subscription, naturally leads to heterogeneity in loyalty or heterogeneity in frailty to terminate the subscription. Thus, we hypothesize:

H6: The frailties are heterogeneous among passengers with different entrance times.

Table 7.2 A summary of the model specifications for the three models

Components	Cox model	Stratified model	Shared frailty model
Joint linear effect of covariates	√	√	√
Moderating effects between factors	√	√	√
Stratified baseline hazard functions		√	
Unobserved heterogeneity			√

The frailty is assumed to be shared by passengers with the same entrance month. It refers to the perceptible service quality in terms of convenience and efficiency in subscribing/returning tickets and information availability.

Thus, to investigate the effects resulting from the unobserved heterogeneity on the service quality in the progress of the CB system, a shared frailty survival model was specified, and passengers were grouped by their first subscription time. The formulation of the shared frailty model is given by:

$$\begin{aligned} h_{ij}(t|\boldsymbol{X_{ij}}) = h_0(t)u_j exp(&\beta_1 F_{ij} + \beta_2 CT_{ij} + \beta_3 OS_{ij} + \beta_4 MT_{ij} + \beta_5 RA_{ij} + \beta_6 TT_{PT-CB,ij} \\ &+\beta_7 TE_{Taxi-CB,ij} + \beta_8 TE_{CB-PT,ij} + \beta_9 ATS_{CB-PT,ij} + \beta_{10} OS_{ij} \cdot F_{ij} \\ &+\beta_{11} RA_{ij} \cdot ATS_{CB-PT,ij}) \end{aligned} \tag{7.14}$$

where j is the index of the cluster grouped depending on the first subscription time, and u_j is the frailty for group j.

Table 7.2 lists the detail of components for all models, and it is clear that the stratified model and shared frailty model are better with deep consideration.

7.5 Results and Discussions

The non-parametric K–M survival curve (see Fig. 7.4) provides an overview of the cumulative proportion of the users retained over time ignoring the effects of any factors. Particularly, the survival rate at each time point t is calculated as follow (Therneau 2020):

$$\widehat{S}(t) = \prod\nolimits_{p:t_p<t}\left(1 - \frac{d_p}{n_p}\right) \tag{7.15}$$

Where $\widehat{S}(t)$ is the estimated survival rate; t_p is the p-th time unit (the time unit in this study is a month); d_p is the number of passengers who terminated the subscription during the time t_p; n_p is the number of passengers who were still active in CB service at the start of time t_p. The 95% confidence interval of survival rate is given by:

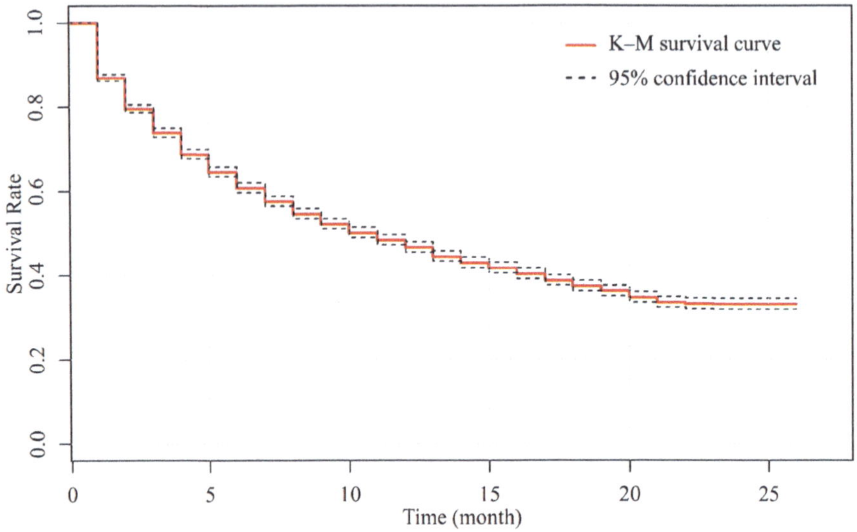

Fig. 7.4 K–M survival curve for CB subscribers

$$CI_{1-\alpha/2} = \widehat{S}(t)^{\exp\left[\pm \frac{Z_{0.025}}{\ln \widehat{S}_{(t)}} \sqrt{\sum_{\{p:tp<t\}} \frac{dp}{np(np-dp)}}\right]} \tag{7.16}$$

Where $Z_{0.025}$ is equal to -1.96. The survival probability initially decreases faster and levels off after approximately 22 months. The results validate our assumption that the first riding experiences are more critical for user loyalty than the repeated ones.

The models built above demonstrated significantly similar parameter estimates. Thus, only the parameter estimates of the shared frailty model are listed in Table 7.3, and a summary of the model performances is presented for comparison in Table 7.4. The Pearson correlation coefficients for the variables included in our models were examined and listed in the appendix (Table 7.5). The correlation coefficients between each pair of explanatory variables are less than 0.5, ensuring the independence of explanatory variables and implying a lower risk of collinearity. All the variance inflation factors (VIFs) reported in Table 7.3 are less than 2, consistently demonstrating that there is not a multicollinearity problem with our model specifications.

The goodness of fit measures for all models are considerably high. This finding indicates that 93–94% of the randomness of the user "survival time" T is explained by the models. The higher partial log-likelihood values at zero indicate that the stratified and shared frailty models show evident improvements in the performances. We believe that the shared frailty model incorporating moderating effects and unobserved heterogeneity in service quality at different stages is more appropriate for modeling and forecasting user loyalty, depending on the measures of the log-likelihood, R^2, and AIC.

Table 7.3 Parameter estimation results for the shared frailty models

Fix effects	Coef	Std. Err	z-Stat	Haz. Ratio	Importance	VIF
Female (F)	0.04	0.052	0.770	1.041	0.001	1.010
Consecutive subscription duration (CT)	−0.527**	0.008	− 67.690	0.591	0.219	1.314
Order of boarding station (OS)	0.022*	0.011	2.120	1.023	0.003	1.059
Maximal interval time (MT)	−1.553**	0.033	−47.390	0.212	0.088	1.310
Rail transit service (RA)	0.278**	0.086	3.240	1.321	0.008	1.551
Differential travel time between PT and CB (TT_{PT-CB})	−0.001	0.001	−1.170	0.999	0.005	1.529
Differential travel expense between taxi and CB ($TE_{Taxi-CB}$)	−0.004*	0.001	−2.510	0.996	0.008	1.731
Differential travel expense between CB and PT (TE_{CB-PT})	0.035**	0.01	3.590	1.036	0.006	1.690
Differential travel speed between CB and PT (ATS_{CB-PT})	0.019**	0.005	3.640	1.020	0.003	1.597
$OS \times F$	−0.033**	0.012	−2.720	0.967		
$RA \times ATS_{CB-PT}$	−0.02**	0.007	−2.990	0.980		
Random effects	Std. dev					
Intercept (grouped by first subscription time, FT)	0.754**					

Note (1) ** 0.01, * 0.05; (2) unstandardized coefficient estimates are reported; (3) the importance denotes the reliable permutation importance (Nicodemus et al. 2010); and (4) the VIF indicates the variance inflation factor calculated using method of (Davis et al. 1986)

Table 7.4 Performances of the survival models

	Cox model	Stratified model	Shared frailty model
Partial log-likelihood at zero	−33,667	−32,510	−32,510
Partial log-likelihood at convergence	−28,376	−27,224	−26,962
R^2	0.93	0.93	0.94
AIC	56,774	54,470	53,871

Sample size 4050 events in 6912 observations

For the characteristics of the users' historical purchase behavior, the statistically significant estimates of the consecutive subscription duration and maximal interval time indicate that the passenger loyalty is influenced by historical behavior characteristics. Previous studies focus on the historical behavior characteristics by including the proxy variables of "occurrence dependence" and "lagged duration dependence" in the survival analysis (Kim and Mannering 1997). Their results also indicate the important correlations between previous behavioral characteristics and the current duration.

The consecutive subscription duration shows a negative effect. Consistent with our assumption H1, this result implies that the longer the period of unceasing subscription for a user, the more loyal he/she is to the CB service subscription. It is the most important predictor for estimating the hazard of subscription termination based on the importance analysis. The hazard ratio is approximately 0.6, which indicates that one month's increase in continuous subscription will strengthen the loyalty of a passenger by approximately 40% (calculated as 1−0.6). This finding implies some valid marketing strategies that will contribute to the reduction in customer churn. For instance, a larger discount for users who have subscribed to the CB service unceasingly for a long time and a longer subscription period option (e.g., quarterly and half-yearly subscriptions) with a time-related quantity discount will contribute to a reduced customer churn.

The maximal interval time also presents a negative effect on the hazard rate. This result indicates that the user who has experienced a longer break and continued to subscribe has a higher loyalty to the CB services, as hypothesized in H2. This variable also makes an important contribution to explaining the randomness in the "survival time" of CB users, second only to the consecutive subscription duration. In particular, a one-month extension in the break interval in the past will result in a loyalty difference of 80%. Thus, the maximum interval time is a highly significant indicator to identify loyal users.

Whether a user has utilized a daily subscription before he/she first joined the monthly subscription also affects his/her loyalty to the CB service. After setting up different baseline hazard functions stratified by the daily subscription experience, the stratified model shows an evident improvement in performance according to the decrease of the AIC and an increase of the log-likelihood in contrast to the Cox model. Figure 7.5 illustrates the adjusted survival curves for users with and without daily subscription experience (results from the stratified model), indicating that the assumption H3 to be correct (i.e. The users who have not utilized the daily subscription show a higher survival rate). It may be because a less flexible long-term subscription increases the pressure to comply with the contract (subscription). Even though the long-term subscription has a significant advantage on price, it also increases the cost of not complying with the contract. There must be some commuters who only need to commute on several specific days of the week. The all-weekdays subscription of monthly ticket increases their cost and therefore naturally weakens their loyalties. Thus, a more flexible subscription strategy, such as subscription on alternate days or other interval-permitted subscription according to the demand, will be beneficial for strengthening user loyalty.

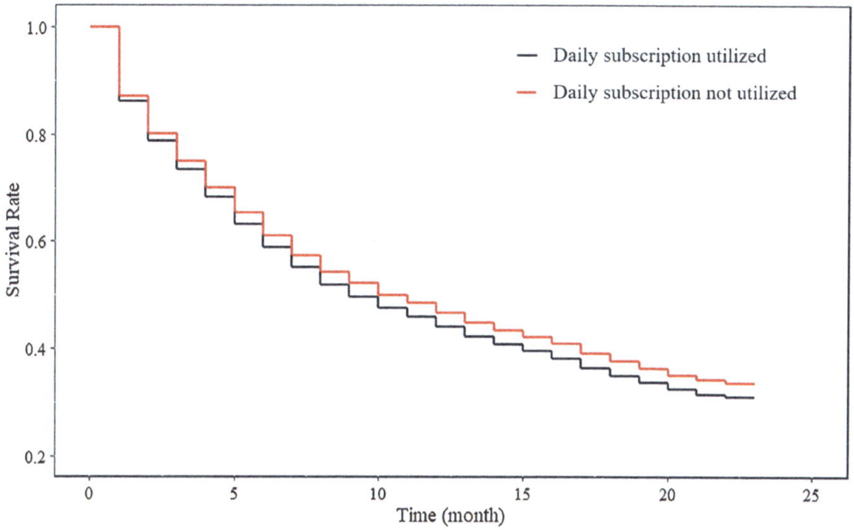

Fig. 7.5 Adjusted survival curves for users with and without daily subscriptions utilized

With the heterogeneous frailties of users included in the shared frailty model, the standard deviation (Std. dev.) in the first subscription time demonstrates that the unobserved heterogeneity of the service quality in the progress of the CB system led to significant variations in user loyalty (verifying the assumption H6). Similar to other covariates, the random frailties can also be understood as the risk score (Therneau 2018). One standard deviation above the mean, i.e., 0.754, corresponds to a relative risk of $\text{Exp}(0.754) = 2.13$, which is more than double the risk of terminating the subscription for users in that group. Thus, neglecting such heterogeneity would result in a biased estimation.

In terms of route and trip-related attributes, most variables included in this study are statistically significant to the hazard of subscription termination. The significant interaction effect between the order of the boarding stations and the female dummy validates our intuition (assumption H4) that females and males have different requirements regarding the order of boarding stations of a CB route. In particular, a unit increase in the order of the boarding stations will produce, on average, a 2.3% increase in hazard of terminating the subscription for males (statistically significant) and a 1.1% decrease (which has been tested as statistically nonsignificant) in that for females. For the passengers who board the CB at the origin station, where there is less uncertainty experienced in the waiting time, females show a lower loyalty toward the CB service than males. However, male loyalty weakens with the number of intermediate stations, and males are ultimately less loyal to the CB service than females. This result may be observed because males are more sensitive to the potentially increased uncertainty in the waiting time caused by increasing intermediate stations and, as a result, more easily shift to other modes of transportation.

The availability of the rail transit service (whether the rail transit services are actually available for passenger's commuting trips) influences user loyalty by changing user perception of his/her usual commute travel speed. Figure 7.6 illustrates the moderating effect between the availability of the rail transit service and the differential travel speed using the coefficients estimated in the shared frailty model and other variables are fixed at their average levels. Essentially, the travel speed of the CB is higher than that of conventional buses but is less than that of rail transit. The positive estimate of differential travel speed articulates that passengers will not become loyal to the CB service for its higher speed in contrast to conventional PT. Thus, the hazard induced by the differential travel speed can be understood as the resistance to accept a change in the speed caused by shifting from PT to CB.

The significant interaction term validates our intuition (assumption H5) regarding the rail transit service in a user's trip which may lead to heterogeneity in satisfaction about the travel speed of the CB service in contrast to PT. In particular, users show a decreasing loyalty with an increase in the differential travel speed if only bus transit service is available. In contrast, the effect of the differential travel speed has been tested as nonsignificant on loyalties for passengers who have rail transit service available for their trips. This may be observed because rail transit riding experiences have made the commuters accustomed to a higher travel speed. Therefore, they can naturally adapt themselves to CB trips which are generally faster than PT trips.

At approximately 14 km/h, the hazard ratio curves intersect with one another, demonstrating a threshold where the levels of loyalty of users in the two user groups are reversed. The rail transit service decreases user loyalty only when the differential travel speed is less than 14 km/h. Owing to the higher speed accustomed, the passengers in "rail transit available" group are more loyal to a much higher travel speed (when the difference is larger than 14 km/h) by the CB.

The differential travel time between PT and CB is of substantial importance in explaining passenger loyalty (Imaz et al. 2014; Trépanier et al. 2012) and is therefore retained in the models, even though it is not significant in the frailty model. As expected, it shows a negative effect on the hazard function, indicating that a shorter travel time makes users more loyal to the CB service. Passengers in the "bus-only group" prefer the CB attributes of reduced travel time but not an increased travel speed. Thus, optimizing the route design to minimize the distance of the detour and travel time does contribute to strengthening their loyalties.

The average cost of a CB trip is higher than that of PT trip but is much cheaper than that of a taxi (Wang et al. 2019a). As expected, the differences in the travel expense between a taxi and CB ($TE_{Taxi-CB}$) and between a CB and PT (TE_{CB-PT}) show different directions but consistent effects on the hazard function, indicating that a cheaper price of subscription enhances user loyalty.

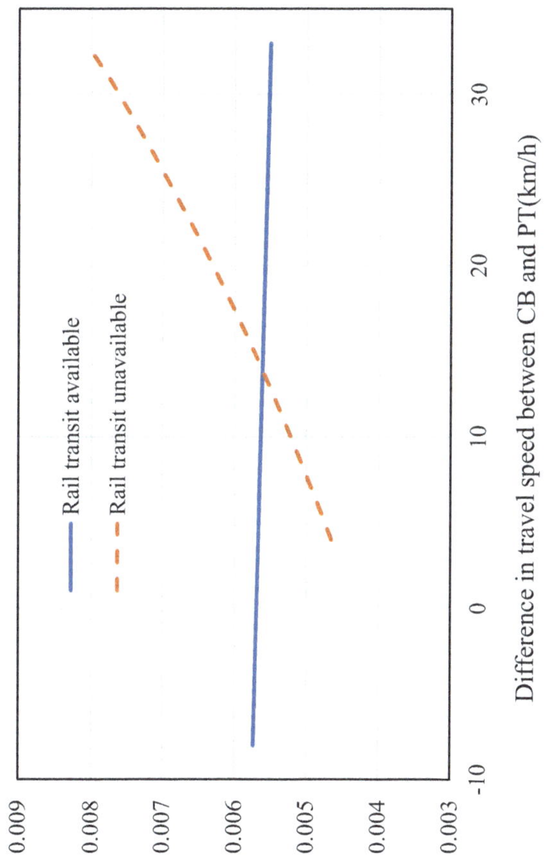

Fig. 7.6 Interaction effect between the availability of the rail transit service and the differential travel speed

7.6 Conclusions

To fill the research gap in user loyalty for the CB service, survival models were employed to determine user loyalty and ascertain key determinants and heterogeneous frailties based on long-term subscription records as well as OD information on CB users in Dalian, China. The use of the survival models enabled analyzing factors that mainly drive user loyalty and facilitated investigating the mechanism of continuous subscription behaviors. The key findings are concluded as follows:

The factors related to the characteristics of the historical purchase behavior, differences in travel attributes in comparison with other available modes, and individual characteristics significantly influence user loyalty in the CB system. The historical purchase characteristics are the most important to user loyalty modeling and forecasting.

The heterogeneous frailties in users significantly contribute to loyalty deviation. They result from the heterogeneity of perceptible service quality in terms of convenience and efficiency in subscribing/returning tickets, as well as information availability in the progression of the CB system.

Male passengers are more sensitive than female passengers to the number of intermediate CB stations due to the potentially increased uncertainty in the waiting time.

The longer the period of unceasing subscription for a user, the more loyal he/she is to the CB service. Thus, a larger discount for users who have subscribed to the CB service unceasingly for a long time and a longer subscription period option (e.g., quarterly and half-yearly subscriptions) with a time-related quantity discount will contribute to a reduction in customer churn.

Passengers without rail transit service available in their trips prefer the CB attributes of a reduced travel time but not an increased travel speed. Thus, optimizing the route design to minimize the distance of detours and the travel time contributes to strengthening their loyalties.

The rail transit service decreases user loyalty only when the differential travel speed is less than 14 km/h. This may be observed because rail transit commuting experiences lead to a higher accustomed travel speed in commuters; therefore, they are more loyal to a much higher travel speed (when the difference is larger than 14 km/h) by the CB.

The experience that a passenger utilized the more flexible daily tickets before subscribing to the monthly ticket for the first time decreases his/her loyalty to the CB service. A more flexible service subscription strategy will be beneficial for strengthening user loyalty, such as a weekly subscription and a subscription on several specific days of a week.

The survival model proposed and calibrated to investigate the loyalty of CB users in this work was demonstrated to be well fitted and robust. The results of the study provide policy-making references for CB operators and governments when developing or improving this type of demand responsive CB system.

This study proposes a fundamental methodology for investigating and modeling the evolution of demand for CB systems. The survival models developed in this study will help CB operators around the world identify and quantify the risk factors due to the flexible nature of such a demand-responsive transit system. This is beneficial for understanding the seemingly irrational behaviors from the aspect of the utility maximization theory. Furthermore, the results of our models will provide references for designing new CB routes and optimizing old routes by minimizing the subscription termination hazards of users.

Some extensions may be of interest to overcome the limitations of this study. Only the gender was examined as an individual characteristic in this study. Other sociodemographic attributes, such as age, income, and private car ownership, may also indicate their effects on loyalty and improve the estimation of the models. In addition, due to the lack of information on the underlying hazard distribution, the study naturally employs the Cox semiparametric models to investigate user loyalty. Some studies indicate that the Cox semiparametric model leads to inefficient parameter estimates, despite the loss in efficiency typically being small (Kim et al. 1993). Therefore, it is worth trying parametric methods after the underlying distributions of the subscription duration are justified (Lee and Wang 2003).

Appendix

See Table 7.5

Table 7.5 Pearson correlation coefficients for variables

	T	F	CT	OS	MT	RA	TT_{PT-CB}	$TE_{Taxi-CB}$	TE_{CB-PT}	ATS_{CB-PT}
Survival time (T)	1									
Female (F)	0.08**	1								
Consecutive subscription duration (CT)	0.776**	0.054**	1							
Order of boarding station (OS)	0.055**	0.043**	0.078**	1						
Maximal interval time (MT)	0.487**	0.061**	−0.057**	−0.01	1					
Rail transit service (RA)	0.016	−0.035**	0.037**	0.043**	−0.016	1				
Differential travel time between PT and CB (TT_{PT-CB})	0.07**	0.029*	0.047**	0.133**	0.051**	0.061**	1			
Differential travel expense between Taxi and CB ($TE_{Taxi-CB}$)	0.005	−0.029*	0.007	0.025*	0.011	0.482**	0.272**	1		
Differential travel expense between CB and PT (TE_{CB-PT})	−0.009	−0.055**	−0.004	−0.131**	0.007	0.377**	−0.002	0.442**	1	
Differential travel speed between CB and PT (ATS_{CB-PT})	−0.037**	−0.016	−0.046**	0.073**	0.01	−0.219**	0.427**	−0.08**	0.159**	1

Note ** 0.01, * 0.05

References

Aiken, L.S., S.G. West, and R.R. Reno. 1991. *Multiple regression: Testing and interpreting interactions*. Newbury Park, California: Sage.

Akaike, H. 1998. Information theory and an extension of the maximum likelihood principle. In *Selected papers of hirotugu akaike*, ed. E. Parzen, K. Tanabe, K.G. 199–213. Springer.

Allenby, G.M., R.P. Leone, and L. Jen. 1999. A dynamic model of purchase timing with application to direct marketing. *Journal of the American Statistical Association* 94: 365–374. https://doi.org/10.1080/01621459.1999.10474127.

Bass, P., P. Donoso, and M. Munizaga. 2011. A model to assess public transport demand stability. *Transportation Research Part a Policy and Practice* 45: 755–764. https://doi.org/10.1016/j.tra.2011.06.003.

Benvenuti, F., C. Diamantini, D. Potena, and E. Storti. 2017. An ontology-based framework to support performance monitoring in public transport systems. *Transportation Research Part C Emerging Technologies* 81: 188–208. https://doi.org/10.1016/j.trc.2017.06.001.

Chi, C.G.Q. 2012. An examination of destination loyalty: Differences between first-time and repeat visitors. *Journal of Hospitality and Tourism Research* 36: 3–24. https://doi.org/10.1177/1096348010382235.

Chu, F. 2015. Customized bus: Cherish what you have at the moment. *Transport Business China* 8: 52–55.

Chung, Y. 2010. Development of an accident duration prediction model on the Korean freeway systems. *Accident Analysis and Prevention* 42: 282–289. https://doi.org/10.1016/J.AAP.2009.08.005.

Cox, D.R. 1972. Regression models and life-tables. *Journal of the Royal Statistical Society: Series B (Methodological)* 34: 187–220. https://doi.org/10.1111/j.2517-6161.1972.tb00899.x.

Cox, D.R., and E.J. Snell. 1989. *The analysis of binary data*, 2nd ed. London: Chapman and Hall/CRC.

Davis, C.E., J.E. Hyde, S.I. Bangdiwala, and J.J. Nelson. 1986. An example of dependencies among variables in a conditional logistic regression. *Modern Statistical Methods in Chronic Disease Epidemiology* 140–147.

Davison, L., M. Enoch, T. Ryley, M. Quddus, and C. Wang. 2012. Identifying potential market niches for demand responsive transport. *Research in Transportation Business and Management* 3: 50–61. https://doi.org/10.1016/j.rtbm.2012.04.007.

DuWors, R.E., and G.H. Haines. 1990. History loyalty analysis measures. *Journal of Marketing Research* 27: 485–493.

Enoch, M., S. Potter., G. Parkhurst, and M. Smith. 2004. Intermode: Innovations in demand responsive transport, final report. Department for Transport and Greater Machester Passenger Transport Executive, London.

Figler, S., P. Sriraj, E. Welch, and N. Yavuz. 2011. Customer loyalty and Chicago, illinois, transit authority buses. *Transportation Research Record: Journal of the Transportation Research Board* 2216: 148–156. https://doi.org/10.3141/2216-16.

Ghasemzadeh, M., B.C.M. Fung, R. Chen, and A. Awasthi. 2014. Anonymizing trajectory data for passenger flow analysis. *Transportation Research Part C Emerging Technologies* 39: 63–79. https://doi.org/10.1016/j.trc.2013.12.003.

Guillén, M., J.P. Nielsen, T.H. Scheike, and A.M. Pérez-Marín. 2012. Time-varying effects in the analysis of customer loyalty: A case study in insurance. *Expert Systems With Applications* 39: 3551–3558. https://doi.org/10.1016/j.eswa.2011.09.045.

Imaz, A., K.M. Nurul Habib, A. Shalaby, and A.O. Idris. 2014. Investigating the factors affecting transit user loyalty. *Public Transport* 7: 39–60. https://doi.org/10.1007/s12469-014-0088-x.

Jacoby, J., and R.W. Chestnut (eds.) 1978. *Brand loyalty: Measurement and management*. New York: John Wiley & Sons.

Jin, L., and L. Guo. 2018. Customer loyalty research based on survival analysis, In *Proceedings—2017 16th international symposium on distributed computing and applications to business, engineering and science, DCABES 2017*, 177–179. https://doi.org/10.1109/DCABES.2017.45.

Jomnonkwao, S., V. Ratanavaraha, B. Khampirat, S. Meeyai, and D. Watthanaklang. 2015. Factors influencing customer loyalty to educational tour buses and measurement invariance across urban and rural zones. *Transportmetrica A: Transport Science* 11: 659–685. https://doi.org/10.1080/23249935.2015.1060274.

Ke, J., H. Zheng, H. Yang, and X.M. Chen. 2017. Short-term forecasting of passenger demand under on-demand ride services: A spatio-temporal deep learning approach. *Transportation Research Part C: Emerging Technologies* 85: 591–608. https://doi.org/10.1016/j.trc.2017.10.016.

Kharoufeh, J.P., and K.G. Goulias. 2002. Nonparametric identification of daily activity durations using kernel density estimators. *Transportation Research Part B: Methodological* 36: 59–82.

Kim, S.G., M. Hamed, and F. Mannering. 1993. *A note on commuters' activity duration and the efficiency of proportional hazards models*. Department of Civil Engineering: University of Washington, Seattle.

Kim, S.G., and F.L. Mannering. 1997. Panel data and activity duration models: Econometric alternatives and applications, In *Panels for Transportation Planning*, 349–373. Springer.

Kleinbaum, D.G., and M. Klein. 2010. *Survival analysis*, 2nd ed. New York: Springer.

Lai, W.T., and C.F. Chen. 2011b. Behavioral intentions of public transit passengers—The roles of service quality, perceived value, satisfaction and involvement. *Transport Policy (Oxf)* 18: 318–325. https://doi.org/10.1016/j.tranpol.2010.09.003.

Lee, E.T., and J.W. Wang. 2003. *Statistical methods for survival data analysis*, 3rd ed. Hoboken, New Jersey: John Wiley & Sons.

Liu, K., J. Wang, T. Yamamoto, and T. Morikawa. 2018. Exploring the interactive effects of ambient temperature and vehicle auxiliary loads on electric vehicle energy consumption. *Applied Energy* 227. https://doi.org/10.1016/j.apenergy.2017.08.074

Liu, K., T. Yamamoto, and T. Morikawa. 2017. Impact of road gradient on energy consumption of electric vehicles. *Transportation Research D Transport and Environment* 54: 74–81.

Liu, T., and A. Ceder. 2015a. Analysis of a new public-transport-service concept: Customized bus in China. *Transport Policy (Oxf)* 39: 63–76. https://doi.org/10.1016/j.tranpol.2015.02.004.

Ma, X. 2015. Why passengers do not accompany with customized bus? *Decision-Making* 10: 82–83.

Mageean, J., and J.D. Nelson. 2003. The evaluation of demand responsive transport services in Europe. *Journal of Transport Geography* 11: 255–270. https://doi.org/10.1016/S0966-6923(03)00026-7.

Mannering, F.L., and M.M. Hamed. 1990. Occurence, frequency, and duration of commuters' work-to-home departure delay. *Transportation Research Part B: Methodological* 24: 99–109.

Marković, N., R. Nair, P. Schonfeld, E. Miller-Hooks, and M. Mohebbi. 2015. Optimizing dial-a-ride services in Maryland: Benefits of computerized routing and scheduling. *Transportation Research Part C Emerging Technologies* 55: 156–165. https://doi.org/10.1016/j.trc.2015.01.011.

Nicodemus, K.K., J.D. Malley, C. Strobl, and A. Ziegler. 2010. The behaviour of random forest permutation-based variable importance measures under predictor correlation. *BMC Bioinformatics* 11: 110.

O'Quigley, J., R. Xu, and J. Stare. 2005. Explained randomness in proportional hazards models. *Statistics in Medicine* 24: 479–489. https://doi.org/10.1002/sim.1946.

Odin, Y., N. Odin, and P. Valette-Florence. 2001. Conceptual and operational aspects of brand loyalty: An empirical investigation. *Journal of Business Research* 53: 75–84. https://doi.org/10.1016/S0148-2963(99)00076-4.

Redman, L., M. Friman, T. Gärling, and T. Hartig. 2013. Quality attributes of public transport that attract car users: A research review. *Transport Policy (Oxf)* 25: 119–127. https://doi.org/10.1016/j.tranpol.2012.11.005.

Therneau, T. 2018. Mixed effects Cox models. *CRAN repository*.

Therneau, T. 2020. A Package for survival analysis in R, *R package version* 3: 2–3.

Tjeendra, M., T.B. BJoewono, and D.U. Ningtyas. 2010. Perceptions and experiences regarding service quality in Indonesian public transport based on gender, In *12th world conference on transport research*, Lisbon, Portugal. 11–15.

Tong, Lu Carol, L. Zhou, J. Liu, X. Zhou. 2017a. Customized bus service design for jointly optimizing passenger-to-vehicle assignment and vehicle routing. *Transportation Research Part C Emerging Technologies* 85: 451–475. https://doi.org/10.1016/j.trc.2017.09.022.

Trépanier, M., K.M.N. Habib, and C. Morency. 2012. Are transit users loyal? Revelations from a hazard model based on smart card data. *Canadian Journal of Civil Engineering* 39: 610–618. https://doi.org/10.1139/l2012-048.

Vatcheva, K., M. Lee, J. McCormick, and M. Rahbar. 2015. The effect of ignoring statistical interactions in regression analyses conducted in epidemiologic Studies: An example with survival analysis using Cox proportional hazards regression model. *Epidemiology* 06. https://doi.org/10.4172/2161-1165.1000216.

Wang, C., M. Quddus, M. Enoch, T. Ryley and L. Davison. 2015. Exploring the propensity to travel by demand responsive transport in the rural area of Lincolnshire in England. *Case Studies on Transport Policy* 3: 129–136. https://doi.org/10.1016/j.cstp.2014.12.006.

Wang, J., T. Yamamoto, and K. Liu. 2019. Role of customized bus services in the transportation system: Insight from actual performance. *Journal of Advanced Transportation*6171532. https://doi.org/10.1155/2019/6171532.

Washington, S.P., M.G. Karlaftis, and F. Mannering. 2003. *Statistical and econometric methods for transportation data analysis*. New York: Chapman and Hall/CRC.

Webb, V. 2010. *Customer loyalty in the public transportation context*. Massachusetts Institute of Technology.

Wen, C.H., L.W. Lan, and H.L. Cheng. 2005. Structural equation modeling to determine passenger loyalty toward intercity bus services. *Journal of Transportation Research Board* 1927: 249–255. https://doi.org/10.3141/1927-28.

Wienke, A. 2011. *Frailty models in survival analysis*. Boca Raton, Florida: Chapman and Hall/CRC.

Yan, N. 2014. A customized bus market survey record [WWW Document]. http://www.chinabuses.com/buses/2014/0729/article_14514.html Accessed 3 Dec 2018.

Part III
Effects of Built Environment on Customized Bus Survival and Demand Evolution

Chapter 8
Effects of Built Environment on Continuous Customized Bus Travel Behavior

Abstract Media coverage of the convenience, flexibility and personalized travel services of demand-responsive transit (DRT) systems masks the fact that these systems have a high failure rate worldwide. The failed cases suggest that DRT does not necessarily survive in all types of human settlements; however, existing studies have failed to give quantitative assessments of how and the degree to which the built environment impacts DRT use. This paper addresses this problem through an empirical study of a successful DRT system in Dalian, China. The results demonstrate a significant impact of the built environment on DRT use. A larger residential population, lower employment density, more residential land uses, less administrative land uses, poor connectivity to road networks and parking supply, better accessibility to different facilities, and long distances from urban business centers encourage DRT use and are therefore conducive to the development of customized bus (CB) systems (routes). The effects of built environment characteristics on DRT use were found to be largely different from those in regular transit systems, indicating that tapping into niche markets underserved by regular transit services is a potential marketing direction for DRT systems. The findings of this study can provide theoretical guidance for DRT operators to identify appropriate market niches and thereby reduce the risk of failure.

Keywords Demand-responsive transit · Built environment · Travel behavior · Customized bus

8.1 Introduction

Demand-responsive transit (DRT) systems are designed to provide transport services on demand from residents, using buses and vans to pick up and drop off passengers in accordance with their personal needs (Currie and Fournier 2020; Enoch et al. 2004; Palmer et al. 2005). Since the 1970s, the increasing desire for convenient, flexible, and individualized public transportation services has given rise to different forms of DRT services and driven their continual improvement (Kirby and Bhatt 1974; Morin 1974).

J. Wang et al., *Customized Bus: User Travel Behavior and Demand Evolution*, Urban Sustainability, https://doi.org/10.1007/978-981-96-3488-0_8

The recent widespread penetration of mobile internet technology in our daily lives has led to the rapid development of these DRT systems, and more systems with internet technology (fundamentally app-based with weak driver contracts and computerized dispatchers) are emerging, providing more people access to customized public travel services, such as "microtransit" in Northern America (Daniel 2017; Jarrett 2019) and "customized bus" ("CB") service in Asia (Abdullah 2017; Liu and Ceder 2015a). However, the hype concerning such systems masks the fact that DRT is very prone to failure (Currie and Fournier 2020). (Currie and Fournier 2020) investigated the performance of 120 DRT schemes worldwide over the last 40 years and found a very high failure rate (with 50% lasting less than 7 years and 40% lasting less than 3 years), especially among recent microtransit systems (with 50% failing within 2 years). CB systems, which recently started in Asia in 2013, also suffer from a risk of easy failure (Chu 2015; Ma 2015; Yan 2014).

A core problem is that due to a lack of prior knowledge as to which types of areas will be susceptible to DRT services (Davison et al. 2012), an introduced DRT operation may not be appropriate for the areas served, leading to dramatic spatial and temporal fluctuations in demand (Jain et al. 2017). Unlike regular bus transit services, DRT systems are commonly designed to provide high-quality and personalized transportation services for a specific group of travelers who have similar travel requests (Enoch et al. 2004; Liu and Ceder 2015a). The travel demands of individuals are assigned as a top priority, and the long-term use behaviors of DRT users strongly affect the ridership of DRT systems (or routes) (Wang et al. 2020b). Particularly, due to high operating costs, DRT systems or routes that cannot maintain sufficient ridership struggle to survive (Currie and Fournier 2020; Enoch et al. 2006).

However, few studies have examined the effects of spatial characteristics on DRT use. Long-term tracking of user behavior and DRT system operation status is essential to identify the effects of built environment characteristics on DRT system survival. Additionally, the lack of accurate and comprehensive measurements of spatial characteristics may be another challenge in clearly identifying the "target market" of DRT services. Based on an empirical study of a successful DRT system in China, this paper aims to (1) reduce the literature gaps through a systematic investigation into the association between the built environment and DRT travel behavior and (2) identify the potential spatial characteristics that encourage DRT use. To achieve the research objectives, the authors collected continuous subscription behavior data of all passengers for more than two years in the Dalian DRT system and comprehensively computed built environment characteristics through multisource data fusion and mining techniques.

The remainder of this paper is structured as follows. Section 8.2 reviews key studies relevant to the investigation of factors that impact DRT use and survival. Detailed descriptions of the study area, data collection, and built environment measurements are given in Sect. 8.3. The model used to model the relationship between the 5 "Ds" and travel behavior is illustrated in Sect. 8.4. Section 8.5 presents the modeling results and a discussion of policy implications. Section 8.6 recapitulates the key findings obtained from the estimation results.

8.2 Literature Review

A series of studies have investigated hundreds of DRT systems across the world and revealed that the sustainable development of DRT systems is very challenging (Currie and Fournier 2020; Davison et al. 2012; Enoch et al. 2006, 2004; Schasché and Sposato 2021). More than half of the studied DRT systems were out of service due to chronic insufficient ridership. One of the major reasons is that failed DRT projects are often designed without a full understanding of the market segments they are serving and therefore cannot maintain ridership over the medium to long term (Davison et al. 2012; Enoch et al. 2006). Three factors have been identified in previous studies to influence DRT use and ridership (Davison et al. 2012; Jain et al. 2017; Wang et al. 2014, 2021): area-related attributes, service supply characteristics, and individual/household characteristics.

The failed cases and successful systems indicate that DRT services are not necessarily viable in all types of human settlements. Travelers will frequently choose the mode of transportation that has higher utility than other available modes. Unlike traditional public transportation systems, DRT systems are demand-driven and user-oriented and are commonly designed to serve specific travel demands, such as commuting trips (Liu and Ceder 2015a). Because of the carpooling nature of DRT services, the travel demands of passengers are highly homogeneous. In fact, the spatial characteristics around passenger neighborhoods have been identified as among the major factors that contribute to the homogeneity of travel demands (Enoch et al. 2006; Wang et al. 2015). Notably, DRT services have been reported to be more likely to survive in areas (i.e., "target areas") with specific spatial characteristics (Currie and Fournier 2020; Enoch et al. 2004).

Area-related factors are commonly measured using a series of built environment variables, as built environment characteristics are the most important indicators of area-related features. A vast literature has demonstrated that built environment characteristics significantly affect the travel behavior of residents, not only for regular transportation modes (Cao et al. 2009; Ewing and Cervero 2010; Zhang et al. 2021, 2020) but also for emerging DRT services (Sabouri et al. 2020). However, systematic examinations of the effects of the built environment on DRT travel behavior are rare in the literature, which may be due to data limitations. Most relevant studies have only qualitatively analyzed the important role of the built environment in DRT market identification through multisystem comparison, and no clear pattern has emerged regarding the directions and magnitudes of the built environment effects on DRT use (Davison et al. 2012; Laws 2009; Scott 2010; Wang et al. 2021).

(Enoch et al. 2006) investigated 72 DRT projects across the world and identified the hostile land use of DRT systems as dispersed land uses with patchy, partially developed areas, areas with widespread cul-de-sacs and winding estate roads. They found that more dispersed land-use patterns generally indicate poor accessibility and therefore induce increased private car ownership (auto travel) and detours in DRT routing. Based on a quantitative analysis of the demand for DRT in the metropolitan region of Greater Manchester, (Wang et al. 2014) confirmed that areas with low car ownership have significantly more DRT demand. In addition, service areas with widespread overly circuitous street structures, such as cul-de-sacs and winding estate roads, lead to long journeys in picking up and dropping off passengers and low driving speed (Currie and Fournier 2020; Enoch et al. 2006, 2004).

Areas with poor regular public transit (PT) services are identified as potential market segments for DRT services (Anspacher et al. 2004; Enoch et al. 2006; Jain et al. 2017; Wang et al. 2021, 2020b). (Potts et al. 2010) found that DRT systems are more likely to survive in areas with large transit-dependent populations and high population densities. Poor PT services can be identified by a lack of public bus/rail-transit station proximity/long walking distances and poor satisfaction with existing services. (Anspacher et al. 2004) investigated DRT services in the San Francisco Bay Area and found that the residents who resided more than half a mile from the nearest transit station were more willing to use the service. A number of successful DRT systems were found to focus mainly on travelers in rural areas where travel demand is low (population density is low) and travel time is long (Currie and Fournier 2020; Scott 2010; Wang et al. 2015). The marketing orientation of these DRT systems should be different from those that have emerged in Asian cities, such as CB systems.

However, there is a lack of quantitative investigations into the effects of built environment characteristics on DRT use in the literature, leading to theoretical gaps in service design, operation optimization, and market identification of DRT systems.

To capture the independent influence of built environment characteristics on DRT travel behavior, we have to control for the effects of the service supply and individual attributes (of users). The quality of service supplied determines the perception of users in using the service and strongly impacts their next choices (Davison et al. 2012; Imaz et al. 2014). The travel time added to the trip to serve other passengers may be a greater barrier to the use of shared services than the presence of a stranger (Lavieri and Bhat 2019). (Chen et al. 2017) found that the travel time is the most important determinant of sharing trips with other passengers. (Alonso-gonzález et al. 2021) pointed out that whether an individual is willing to share his or her ride with other passengers depends mainly on the time–cost trade-offs, such as detours. (Wang et al. 2021) reported a significant negative effect of price rate on DRT demand and found price elasticity at approximately − 0.3, which is similar to that of regular public bus services. Additionally, several studies have identified certain individual or household socioeconomic characteristics that impact the willingness of individuals to choose demand-responsive travel services. For instance, females, young people, and individuals with low vehicle ownership and a low income level are more willing to use these services (Jain et al. 2017; Lavieri and Bhat 2019; Wang et al. 2014). Moreover, travel purposes also affect the use of DRT. (Lavieri and Bhat 2019) observed that users for commuting trips are less sensitive to the presence of strangers than users with other travel purposes. This may be one of the reasons why DRT services are much more popular for commuters than other types of travelers (Ndibatya and Booysen 2020; Schasché and Sposato 2021).

8.3 Study Area and Built Environment Measurements

CB services are a recently emerged type of DRT system (Liu and Ceder 2015a) and have attracted increasing numbers of travelers in some large and medium-sized cities in Asia, such as Beijing (Ma et al. 2017), Shanghai (Li 2019), Dalian (Wang et al. 2019a), and Singapore (Abdullah 2017), since their first operation in Qingdao, China (Zhan and Dong 2014). Despite the rapid increase in popularity of CB services, we can also observe a high failure rate similar to those of other DRT systems (Currie and Fournier 2020); examples include the CB systems in Guangzhou, Hangzhou, and Nanjing, China (Wang et al. 2019a). A successful system with numerous regular passengers can often provide practical experience and lessons for exploring the effects of the built environment on CB system survival. The Dalian CB system, which was established in early 2014 and serves 4000–6000 person-times per day, was chosen as a case study in this paper.

The Dalian CB system is designed mainly for commuting trips. The demands are relatively regular, e.g., demand for travel from a residence to a workplace every weekday. The users typically purchase CB travel services on a monthly basis (monthly subscription). In this study, the subscription records of the Dalian CB system were collected, and it was found that approximately 84% of users subscribed to the services repeatedly and created 99% of orders. Since the onset of normal operation of the Dalian CB system (after Dec. 2015), orders from new users (nonsubscribers) have been very limited. On average, only approximately 8% of the subscribers to CB services each month are new users. Therefore, the continuous subscriptions of the regular subscribers play a decisive role in the survival of CB routes. The subscription frequencies of the subscribers basically determine the ridership of the CB system. The lower the frequency of subscribers' subscriptions, the smaller the ridership and the more difficult it is for the system to survive (maintain sufficient ridership). Thus, this study was aimed at identifying the characteristics of urban areas that are conducive to the development and survival of CB systems by examining the associations between subscriber subscription frequency and built environment characteristics.

8.3.1 Dalian CB System

Dalian is a major coastal city in Northeast China (as illustrated in Fig. 8.1) that plays a leading role in the economic development of Liaoning Province, China, and contains more than 7.45 million people according to the 7th National Population Census of China. Passenger car ownership has reached more than 1.79 million at the end of 2020 according to the Dalian Statistical Yearbook of 2021.

The Dalian CB system mainly serves commuting travel demands. Prospective users can submit their travel demands on a smartphone application and provide the origin, destination, expected departure time, and expected arrival time of their

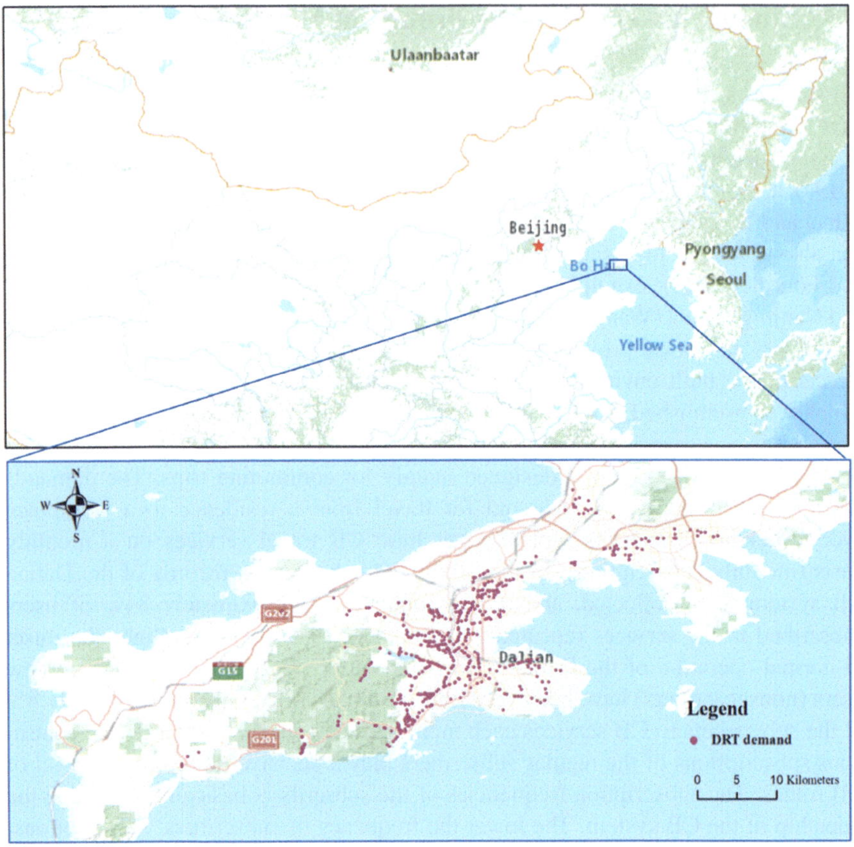

Fig. 8.1 Location of Dalian, China, and the distribution of DRT demand (subscribers' homes)

trip. Upon receiving these travel demands, the CB operators design preliminary CB routes (with origin stops and terminal stops of the CB routes designed) to recruit riders. Travelers can easily choose their preferred routes and make a reservation and payment using the smartphone application or the easy-to-access website via what is commonly referred to as a CB service subscription.

There are two types of subscriptions in the Dalian CB system: daily subscriptions and monthly subscriptions. Monthly subscriptions (which are valid for all weekdays of a specific month) are held by the majority of those requesting service in the Dalian CB system and therefore are the focus of this study. Generally, the price for CB service is approximately two to three times more expensive than regular bus travel but much cheaper than traveling by taxi, at approximately one-third the cost.

As illustrated in Fig. 8.2, the number of monthly subscriptions first exceeded 5000 in Dec. 2015 and remained approximately steady thereafter. Therefore, the steady period from Dec. 2015 to Sep. 2017 was employed as the research period in this study. The monthly subscription records for a total of 7052 subscribers in the research period

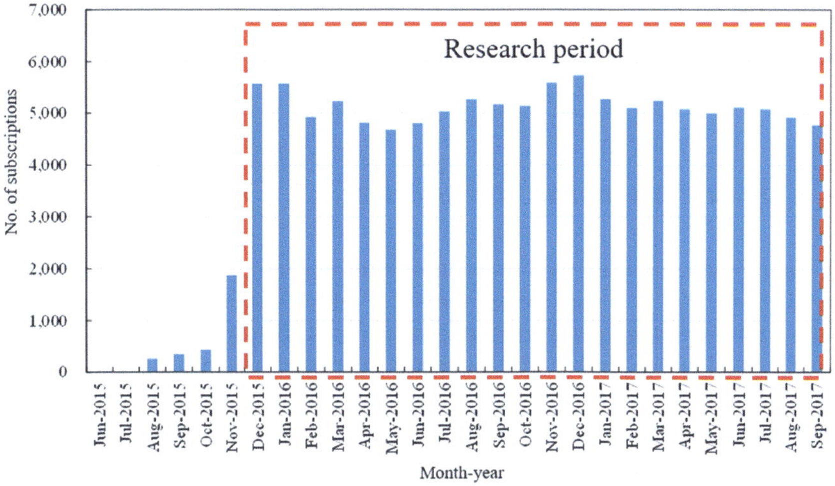

Fig. 8.2 Demand for the Dalian CB system

were collected to investigate the built environment effects. Each subscription corresponds to an actually completed CB order and contains the subscriber ID, location of the boarding/deboarding stop, subscription price, subscription time, and some information on the subscribed CB route.

8.3.2 *Data Collection*

One of the main barriers to investigating the effects of the built environment on travel behavior is the absence of detailed information on built environments. To overcome this challenge, multiscale geographic data and additional open-to-access big data sources were fused to measure the 5 "Ds" of the built environment based on the research framework proposed in Sect. 2.1. The following data were collected in this study:

- 2015 urban land use data, obtained from the Bureau of Natural Resources, Dalian. Eight types of land uses were classified: administrative and public services, commercial and business facilities, green spaces and parks, industrial and manufacturing spaces, residential areas, transportation, municipal utilities, and logistics and warehouses.
- The distributions of population and employment, which were estimated from China Liaoning Unicom cellular location data (including one month of mobile station data in 2018)
- POI data, obtained from Gaode Map (including all types of POIs on Gaode Map mined in 2018)

- Vector geographic data, obtained from Open Street Map (including the driving and walking road network in 2019)
- Subscription records, representing long-term continuous observations (including the coordinates of trip origin/destination, costs, etc.).

8.3.3 Built Environment Measurements

The most commonly used measurement of built environment is based on "5Ds". Although the "5Ds" is mainly created for VMT/VHT or vehicle-oriented studies, the five dimensions in whole or in part have been widely used to capture built environment characteristics and discovered to significantly affect the travel behavior of residents for almost all types of transportation modes, including walking (Yang et al. 2021), bicycling (Zhao et al. 2019), public transit riding (Ding et al. 2014; Ewing and Cervero 2010), and even the DRT (Huo et al. 2021; Sultana et al. 2018) and ride-hailing service use (Sabouri et al. 2020). The built environment impacts travel behavior because the built environment determines whether a travel mode choice is available or easily accessible to an individual (Ewing and Cervero 2010; Handy 2018).

As a complement to the traditional public transportation system, although urban planners may not adjust built environment in planning practice to better suit DRT development, ascertaining the effects of built environment on DRT use can significantly help operators to identify the target market more effectively and reliably, and to reduce the risk of failure. On the basis of the above datasets, elements pertaining to the 5 "Ds" of the built environment, i.e., *density, diversity, design, distance to transit*, and *destination accessibility* (Cervero and Kockelman 1997; Ewing and Cervero 2010), were computed for each CB subscriber.

Regarding the first "D" of the built environment, i.e., *density*, distributions of population and employment are commonly employed in studies as valid indicators (Cervero and Kockelman 1997; Munshi 2016). In this study, the population density within a 1000 m buffer zone of each subscriber's household was calculated based on cellular signaling data from China Liaoning Unicom as an indicator to capture the population distribution. The employment density, as another indicator of *density*, was calculated as follows (Munshi 2016):

$$D_{emp} = \frac{N_w}{A_e} \tag{8.1}$$

where D_{emp} is employment density; N_w denotes the number of workers and was estimated based on cellular signaling data from China Unicom in 2018; and A_e is the area under economic activity, including administrative and public services, commercial and business facilities, and industrial and manufacturing spaces.

Balanced and mixed land use has been advocated to encourage more walking and cycling and less vehicle travel (Cao et al. 2006; Huang and Tsai 2013). The entropy and dissimilarity indices developed by (Cervero and Kockelman 1997) have been widely used to capture the *diversity* dimension of the built environment. The entropy index measures the balance level of land use, where 1 signifies perfect balance and 0 denotes only a single type of land use (Munshi 2016; Sehatzadeh et al. 2011). We calculated the entropy index for each sampled subscriber with the formulation given by:

$$E = -\sum_{j=1}^{8} \frac{P_j \times ln(P_j)}{ln(J)} \tag{8.2}$$

where E is the entropy index for the 1000 m buffer area around a subscriber's household; P_j is the proportion of the floor space area in the j-th type of land use, and $J = 8$, which is the total number of land-use types considered in this study.

The dissimilarity index is another indicator of land use mix and measures the degree to which the land-use types abutting or diagonal to each hectare differ. The dissimilarity index was computed based on points awarded to a central cell (as shown in Fig. 8.3):

$$DI = \frac{1}{K} \sum_{i=1}^{K} \sum_{l=1}^{8} \frac{PO_{il}}{8} \tag{8.3}$$

where DI is the dissimilarity index for a hectare grid, K is the total number of grids enclosed within the 1000 m buffer around the location of a subscriber's household, and PO is the points awarded for a cell enclosed by the 1000 m buffer.

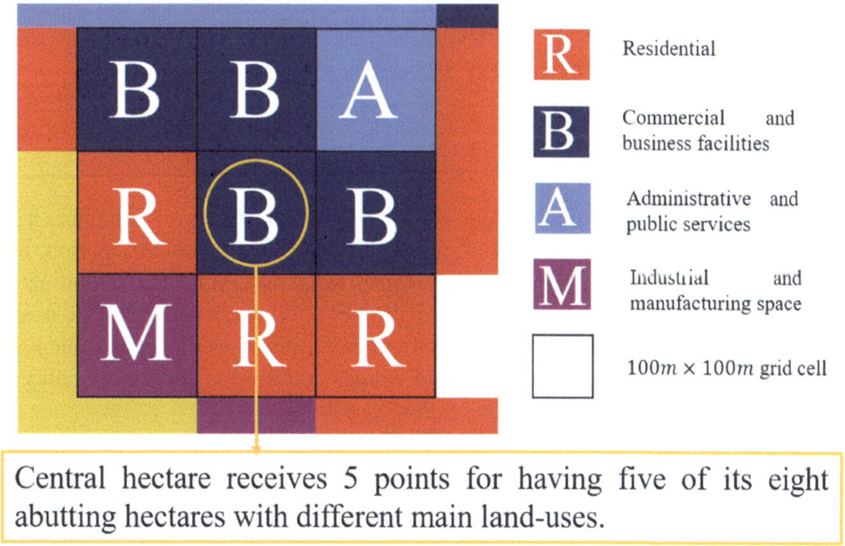

Fig. 8.3 Computation of the dissimilarity index (the study area is divided by a fishnet with $100m \times 100m$ grid cell)

The third "D" (*design*) measures the road/street network characteristics, pedestrian and cycling provision, and site design within a given area (Cervero and Kockelman 1997; Ewing and Cervero 2010). Connectivity to the road network was employed as an indicator of road/street network characteristics, while the number of parking lots was considered a proxy quantifying the site design (Cervero and Kockelman 1997; Yang et al. 2019). The connectivity to the road network and the number of parking lots within the 1000 m buffer of each household were assessed based on the Dalian POI data. Specifically, the total number of roundabouts, intersections, and expressway entrances and exits in the road network within the buffer of each household were counted and summed as the connectivity indicator (Ewing and Cervero 2010; Frank et al. 2004).

The fourth "D" is the *distance to transit services* (Munshi 2016; Ogra and Ndebele 2013), which captures the proximity and accessibility of transit services. The Euclidean distance from each household to the nearest bus stop and rail transit station were calculated as indicators of the fourth dimension (Kitamura and Mokhtarian 1997).

Destination accessibility measures the ease of access to trip attractions (Ewing and Cervero 2010). Distance to business centers and distance from different types of facilities have often been used as indicators in previous studies (Cao 2010; Ding et al. 2018b; Guan and Wang 2019a; Yang and Cao 2018). The driving path distance in the Dalian road network from each household to the 13 business centers (including Qingniwa, Tianjinjie, Xianlu, Heping, Heishijiao, Wandagaoxin, Aolinpike, Dongte, Huananguanchang, Ganjingzilu, Quanshui, Bayilu, and Kaifaquansheng) and the walking distance from each household to the nearest 8 types of land use were calculated and examined in the models.

8.3.4 Other Factors

To investigate the effects of the built environment on DRT travel behavior, it is necessary to control for other influencing factors. During the research period, the Dalian government did not provide any subsidies or investments in the operation of DRT systems. Therefore, the impact of government policy changes on DRT usage behavior was ignored in this study. To address other aspects, we included indicators of sociodemographic characteristics of individuals and service supply attributes to control the estimation bias caused by omitted variables.

Gender and housing price were examined in this study. Due to privacy issues, income information for passengers is not available. We obtained the transaction price data of second-hand houses in Dalian and calculated the average housing price for each passenger, which is partially indicative of income level. Additionally, according to an in-person interview with the service operator, most passengers are young commuters between the ages of 20 and 45. Therefore, the impact of an overly concentrated age distribution on the use of DRT is limited in the Dalian DRT system. The monthly subscription price per kilometer was calculated based on the subscription records to capture the supply characteristic of the DRT services.

The variables included in this study are summarized in Table 8.1.

8.4 Modeling of DRT Travel Behavior

8.4.1 Random-Effects Negative Binomial Regression Model

As frequency data are always nonnegative integer values, count data models are potentially applicable. As evident from the mean and standard deviation of each response variable in Table 8.1, e.g., a mean of 7.84 and a standard deviation corresponding to a variance of 61.47, the variance is much larger than the mean. Therefore, the negative binomial model, which allows for overdispersion in responses, was appropriate for the dataset.

A previous study indicated that the heterogeneity of service quality the first time a user purchases a monthly subscription contributes significantly to heterogeneity in subscription behavior (Wang et al. 2020b). To control for this natural heterogeneity, a random-effects negative binomial regression (RNB) model was employed to model the relationship between subscription behavior and the built environment.

The number of occurrences (counts) of an event is commonly assumed to follow a Poisson distribution and be conditional on the latent variable u_{ij}^* and random effects $\boldsymbol{v}_j$, given a series of covariates (StataCorp 2017):

$$y_{ij}|\mu_{ij}^* \sim Poisson\left(\mu_{ij}^*\right) \tag{8.4}$$

$$\mu_{ij}^*|\boldsymbol{v}_j \sim Gamma\left(r_{ij}, p_{ij}\right) \tag{8.5}$$

$$\boldsymbol{v}_j \sim N\left(0, \sigma^2\right) \tag{8.6}$$

where y_{ij} is the count response of the i-th observation from the j-th group ($i = 1, \ldots, n_j, j = 1, \ldots, M$); r_{ij} and p_{ij} are the shape parameter and scale parameter, respectively; and $\boldsymbol{v}_j$ is the random effects and is assumed to follow a normal distribution with mean 0 and variance σ^2.

Therefore, the conditional probability of observing $y_{ij} = y$ is given by:

$$\Pr\left(y_{ij} = y|\boldsymbol{v}_j\right) = \frac{\Gamma\left(y+r_{ij}\right)}{\Gamma(y+1)\Gamma\left(r_{ij}\right)} p_{ij}^{r_{ij}} \left(1 - p_{ij}\right)^y \tag{8.7}$$

The expectation is formulated as:

$$E\left(y_{ij}|\boldsymbol{X}_{ij}, \boldsymbol{v}_j\right) = \mu_{ij} = \exp\left(\boldsymbol{X}_{ij}\boldsymbol{\beta} + \boldsymbol{v}_j\right) \tag{8.8}$$

Table 8.1 Descriptions and descriptive statistics of variables (sample size = 7052)

Variable	Description	Mean	S.D.
Frequency of subscription	Subscription frequency during the research period	7.84	6.62
Sociodemographic characteristics			
Gender (Female)	Coded as 1 for females and 0 for males	0.73	0.44
Housing price	Housing price per square meter around a subscriber's household (10 thousand RMB per square meter)	0.99	0.15
Service supply attribute			
Subscription price	Monthly subscription price per kilometer (100 RMB per km per month)	0.14	0.05
Density			
Population density 2018	Residential population density (in 2018) within the 1000 m buffer around a subscriber's household, divided by 100,000 people	0.04	0.03
Employment density 2018	Employment density (in 2018) within the 1000 m buffer around a subscriber's household (100,000 people per square kilometer)	0.15	0.43
Diversity			
Entropy index	The entropy index within the 1000 m buffer around a subscriber's household	0.49	0.18
Entropy index	The dissimilarity index within the 1000 m buffer around a subscriber's household	0.30	0.07
Design			
Connectivity	Number of nodes connected to the road network within the 1000 m buffer around a subscriber's household, divided by 100	0.36	0.50
Parking lots	Number of parking lots within the 1000 m buffer around a subscriber's household, divided by 100	0.47	0.50
Distance to transit			
Distance to rail station	Euclidean distance from a household to its nearest rail station (km)	1.53	1.30
Distance to bus stop	Euclidean distance from a household to its nearest bus stop (km)	0.22	0.30

(continued)

Table 8.1 (continued)

Variable	Description	Mean	S.D.
Destination accessibility			
Distance to 8 land-use types	Average walking distance from the nearest 8 types of land use (km)	1.33	0.88
Distance to business center	Shortest path distance in Dalian road network from a household to the nearest business center (km)	5.12	3.38
Variable	Description	Mean	S.D.
Land use proportion			
Administrative proportion	Proportion of land use for administrative and public services within the 1000 m buffer around a subscriber's household	0.09	0.10
Commercial proportion	Proportion of land use for commercial and business facilities within the 1000 m buffer around a subscriber's household	0.05	0.08
Green-space proportion	Proportion of land use for green spaces and parks within the 1000 m buffer around a subscriber's household	0.05	0.07
Industrial proportion	Proportion of land use for industrial and manufacturing spaces within the 1000 m buffer around a subscriber's household	0.22	0.22
Residential proportion	Proportion of land use for residential use within the 1000 m buffer around a subscriber's household	0.55	0.22
Transportation proportion	Proportion of land use for transportation facilities within the 1000 m buffer around a subscriber's household	0.01	0.02
Municipal proportion	Proportion of land use for municipal utilities within the 1000 m buffer around a subscriber's household	0.02	0.03
Logistics and warehouse proportion	Proportion of land use for logistics and warehouse utilities within the 1000 m buffer around a subscriber's household	0	0

where $\boldsymbol{X}_{ij}$ is the row vector of the fixed-effects covariates with the fixed regression coefficients $\boldsymbol{\beta}$ and $\boldsymbol{v}_j$ is the vector of the random effects (with only the random intercepts included in this study).

In this study, mean-conditional-overdispersion parameterization was employed as follows:

$$r_{ij} = 1/\alpha \; and \; p_{ij} = \frac{1}{1+\alpha\mu_{ij}} \tag{8.9}$$

Therefore, the conditional overdispersion is $1 + \alpha\mu_{ij}$ (Carmeron and Trivedi 1986), where α is regarded as a conditional overdispersion parameter. The marginal dispersion can be calculated by:

$$Var(y_{ij}) = 1 + \left[\exp(\sigma^2)(1+\alpha) - 1\right]E(y_{ij})E(y_{ij}) \tag{8.10}$$

where the values of σ^2 and α jointly determine the level of overdispersion.

Given the random effects $\boldsymbol{v}_j$ and the conditional overdispersion parameter α, the conditional distribution of $\boldsymbol{y}_j = \left(y_{1j}, \ldots, y_{n_j j}\right)^T$ is given by:

$$\begin{aligned} f\left(\boldsymbol{y}_j|\boldsymbol{v}_j, \alpha\right) &= \prod_{i=1}^{n_j} \tfrac{\Gamma(y_{ij}+r_{ij})}{\Gamma(y_{ij}+1)\Gamma(r_{ij})} p_{ij}^{r_{ij}} \left(1 - p_{ij}\right)^{y_{ij}} \\ &= exp(\sum_{i=1}^{n_j}\{log\Gamma\left(y_{ij} + \tfrac{1}{\alpha}\right) - log\Gamma\left(y_{ij} + 1\right) - log\Gamma\left(\tfrac{1}{\alpha}\right) \\ -\tfrac{1}{\alpha}\log&\left[1 + \exp\left(\boldsymbol{X}_{ij}\boldsymbol{\beta} + \boldsymbol{v}_j + log\alpha\right)\right] - y_{ij}log\left[1 + exp\left(-\boldsymbol{X}_{ij}\beta - \boldsymbol{v}_j - log\alpha\right)\right]\}) \end{aligned} \tag{8.11}$$

where r_{ij}, p_{ij}, and μ_{ij} are the same as in Eqs. (8.8) and (8.9).

8.4.2 Model Specification

The factors related to the individual sociodemographic, service supply, and five dimensions of built environment characteristics around each user's household were examined using a series of RNB models. We computed multiple variables for each dimension to comprehensively capture the potential characteristics of the built environment. However, this process may lead to multicollinearity when all variables are incorporated into one function simultaneously. For instance, the entropy index and dissimilarity index were identified as correlated, with a Pearson correlation coefficient of 0.73, significant at the 0.01 level; connectivity and number of parking lots had a Pearson correlation coefficient of 0.80, significant at the 0.01 level; and distance to the rail station and distance to the bus stop had a Pearson correlation coefficient of 0.42, significant at the 0.01 level. We specified Model 1 and Model 2 to examine these pairs of variables separately to verify the effects of the related dimensions of the built environment and validate the robustness of the model specification. Additionally, since one of our research objectives was to determine the market direction of CB services, we constructed models (Models 3 and 4) to investigate what types of land use are conducive to the development and survival of CB systems.

As indicated by previous studies (Chi 2012; Wang et al. 2020b), satisfaction is more significant for first-time experiences than repeated experiences. Thus, the heterogeneity in satisfaction with the first time a user purchases a monthly subscription may lead to heterogeneity in subscription behavior. Therefore, a random intercept was specified for all of the models to control for this heterogeneity. Passengers were grouped by their first month as CB users, and users in the same group shared the same random intercept. This approach allowed us to focus on the effects of other factors at different stages of CB development on the CB travel behavior of users without the interference of heterogeneity in service quality.

8.5 Results and Discussion

All the models were estimated using STATA 15.1, and the estimation results are summarized in Table 8.2. The significant overdispersion parameter $log\alpha$ in the models indicates that the negative binomial regression model fits the data better than Poisson regression. The heterogeneity in service quality correlated with (grouped by) the first subscription month was found to be highly significant, indicating an appropriate model specification. The coefficient estimates present similar magnitudes and directions among the different models, demonstrating the robustness of the model specification.

The estimation results indicate that multiple dimensions of built environment variables have significant impacts on CB use. Regarding the *density* dimension, residential population density positively affects the use of CB services. Generally, the density of the transit-dependent population is proportional to the residential population density, and the demand for DRT systems highly depends on the transit-dependent population and its density (Potts et al. 2010). Due to the high operating costs of CB systems, sufficient ridership is a prerequisite for the sustainable operation of CB routes. Once a CB route is chronically underattended (has insufficient users subscribing to the route), it will be canceled. It is easier to maintain a high ridership for a CB route in a neighborhood with a high residential population density. Employment density has a significant negative effect on CB use. Users who reside in areas with fewer employment opportunities are more likely to frequently subscribe to CB services. This relationship may be observed because areas with larger residential populations and lower employment densities have greater outbound commuting demands. Therefore, CB routes in such areas are more likely to have an adequate and stable customer base (passengers with high frequencies of use) and therefore are more likely to survive.

Regarding the indicators of land-use mix, both the entropy and dissimilarity indices present significant negative effects, indicating that highly mixed land-use patterns are not conducive to frequent CB use. This finding is inconsistent with results for conventional transit systems (Ewing and Cervero 2010; Frank et al. 2008), which may imply a difference in market orientations between CB and regular bus services. CB services are mainly designed for specific types of travel demand (Liu and Ceder 2015), and such homogeneous travel demand mostly arises in areas with similar land uses. To further explore what type of land use encourages CB use, the proportions of the eight types of land use were examined with the models excluding land-use mix factors to avoid multicollinearity. Among the land-use types, residential and administrative land uses were found to have the strongest effects on CB subscription frequency (see the estimation results of Model 3 and Model 4 shown in Table 8.2). As

Table 8.2 Effects of the built environment on CB subscription behavior

	Model 1	Model 2	Model 3	Model 4
	Coeff. (z-stat)	**Coeff. (z-stat)**	**Coeff. (z-stat)**	**Coeff. (z-stat)**
Sociodemographic				
Gender (Female)	0.110***	0.112***	0.108***	0.107***
	(5.19)	(5.26)	(5.10)	(5.03)
Housing price per square meter (10 thousand RMB per square meter)	0.080	0.087	0.069	0.192**
	(0.86)	(0.99)	(0.74)	(2.10)
Service supply feature				
Subscription price (100 RMB per km per month)	−0.854***	−0.870***	−0.792***	−0.720***
	(−3.90)	(−3.90)	(−3.61)	(−3.24)
Built environment characteristics				
Density				
Population density 2018	2.715***	2.460***	2.494***	2.465***
	(4.78)	(4.13)	(4.27)	(4.31)
Employment density 2018	−0.073***	−0.070***	−0.067***	−0.068***
	(−3.01)	(−2.99)	(−2.69)	(−2.88)
Diversity				
Entropy index	−0.168**			
	(−2.45)			
Dissimilarity index		−0.487***		
		(−2.81)		
	Model 1	Model 2	Model 3	Model 4
	Coeff. (z-stat)	**Coeff. (z-stat)**	**Coeff. (z-stat)**	**Coeff. (z-stat)**
Design				
Connectivity		− 0.027		
		(−1.01)		
Parking lots	− 0.078***		−0.077***	−0.087***
	(−3.06)		(−2.92)	(−3.52)
Distance to transit				
Distance to rail station		0.012	0.009	0.013
		(1.40)	(1.12)	(1.51)
Distance to bus stop	0.013			
	(0.35)			
Distance to 8 land-use types	−0.101***	−0.100***	−0.094***	−0.089***
	(−6.68)	(−6.77)	(−6.53)	(−6.12)
Distance to business center	0.013***	0.014***	0.014***	0.012***
	(3.50)	(3.85)	(3.66)	(3.17)

(continued)

Table 8.2 (continued)

	Model 1	Model 2	Model 3	Model 4
	Coeff. (z-stat)	**Coeff. (z-stat)**	**Coeff. (z-stat)**	**Coeff. (z-stat)**
Land use proportion				
Residential proportion			0.067	
			(1.25)	
Administrative proportion				−0.289**
				(−2.47)
Constant	1.684***	1.701***	1.551***	1.487***
	(11.11)	(10.97)	(11.22)	(10.64)
logα	− 0.798***	− 0.796***	− 0.797***	−0.798***
	(−35.37)	(−35.32)	(−35.35)	(−35.37)
Random effect				
Var (first subscription month)	0.301***	0.303***	0.302***	0.301***
	(3.53)	(3.54)	(3.54)	(3.53)
Log-likelihood	−20,712	− 20,715	−20,713	−20,711
N	7052	7052	7052	7052

Note *** $p < 0.01$; ** $p < 0.05$; * $p < 0.1$

intuited, residential land use positively affects CB use. This relationship may have been observed because the CB system mainly serves commuting trips, transporting users between their homes and workplaces. Urban areas with concentrated residential land use are more conducive to a high ridership for CB routes. Administrative land is where schools, governmental agencies, hospitals, etc., are located, and administrative land use is one of the main land uses providing employment opportunities. As expected, it presents a significantly negative effect on CB travel frequency. Therefore, urban areas with more residential land use and less concentrated administrative land use are more likely to be potential/target markets for DRT services that mainly serve commuting trips.

The negative estimates of the two indicators of *design* demonstrate that urban areas with poorer *design* characteristics are conducive to CB use. Street connectivity has been found to increase regular bus use significantly (Frank et al. 2008), and more convenient parking lots in neighborhoods increase the propensity to drive to commute (Guo 2013; Weinberger 2012). Poor connectivity to the road network and parking supply is not conducive to either the development of regular bus systems or private car travel. CB services are irreplaceable for subscribers who reside in such areas. Therefore, areas with poor connectivity and parking supply may leave larger potential market spaces for zero-interchange, point-to-point CB travel services. Furthermore, (Laws 2009) found that DRT systems are more likely to succeed in areas lacking traditional bus and taxi service adaptability.

In terms of *destination accessibility*, the longer the distance to different types of land uses is, the more dispersed the land-use pattern for a neighborhood is. The model estimates indicate that a more dispersed land-use pattern is not conducive to CB use. (Enoch et al. 2006) investigated 72 DRT projects across the world and identified the dispersed land-use pattern as a characteristic hostile to the development of DRT services. This relationship may be observed because more dispersed land-use patterns induce increased private car ownership and auto travel (Wang et al. 2014) and therefore leave less space for CB. Furthermore, dispersed land-use patterns increase the challenge of CB route and station design. In particular, they may easily result in long detours during passenger pick-up and drop-off and reduce the quality and reliability of CB services.

The distance to business centers has a positive effect on CB use. Subscribers who reside in neighborhoods that are far from business centers are more likely to use the CB service frequently. This phenomenon may arise because urban business centers and their surrounding areas provide major employment opportunities for residents. The further a subscriber lives from the business center, the longer the commute tends to be. This result is consistent with the findings reported in previous studies (Kirby and Bhatt 1975; Zhang et al. 2017). One possible reason for this finding is that CB services have significant advantages in service quality for long-distance trips over traditional PT services, such as shorter travel times and a more comfortable travel experience, given the one-seat-guaranteed point-to-point service (Wang et al. 2019a). Ensuring the competitiveness of DRT services over alternatives has been identified as a key factor of success (Enoch et al. 2004; Wang et al. 2019). For those who drive to commute, rather than enduring the fatigue of long hours of driving, taking a CB ride may be preferable. Another possible reason for the above finding may be that travelers are more tolerant of the response time of DRTs on long-distance trips than on short-distance trips.

As expected, the estimates of the indicators used to capture information regarding the *distance to transit* indicate that a poor service level of transit induces more frequent CB use (although neither coefficient is significant). This finding is consistent with the conclusion of (Wang et al. 2021, 2020b) that areas with poor satisfaction with regular bus services are typical markets for DRT services. The CB system in the current study is in a position to complement regular PT by supplying high-quality and customized commuting transit services in areas with poor or inconvenient PT services, similar to the "network DRT" defined by (Enoch et al. 2004).

Females are more likely to use CB services frequently. More frequent usage by females has also been found in other DRT systems, such as those in some European cities (Mageean and Nelson 2003; Wang et al. 2015). This pattern may arise because the vehicle ownership rate of women is lower than that of men (Department Statista Research 2016), with traditional transit and similar transit services being their dominant modes of commuting. In the present study, female subscribers subscribed to the CB service 12% more often than male users ($e^{0.11} \approx 1.12$) given other variables controlled at constant values.

The coefficient estimate of housing price is significant in Model 4 and is positive. This finding indicates that subscribers with higher affordability (who reside in urban areas with high housing prices) are more likely to use CB frequently. This result reflects the difference in service orientation between the CB system and regular bus transportation services: the CB system aims to provide high-quality bus transportation services for some specific types of travel demand, while regular bus services are designed for general travel demand (Balcombe et al. 2004; Liu et al. 2016b; Wang et al. 2019a).

As intuited, subscription price has a significantly negative effect (−0.889 to −0.720) on the frequency of use. The marginal effect of subscription price on the frequency of use is very small; for every 10% decrease in price, the average monthly subscription increases by only 0.02%. For every 11.2–13.9 RMB reduction per km per month, the average monthly subscription increases by 0.1 times. (Currie and Fournier 2020) investigated 120 DRT systems worldwide and consistently found a definitive positive association between price and DRT failure. Emphasis on long-distance trips is conducive to the survival and development of CB systems because operators generally provide discounts for long-distance trips to decrease the service price and therefore attract potential commuters.

8.6 Conclusions

This paper demonstrates the significant impacts of the built environment on the long-term stability of DRT demand, which is regarded as a precondition for the sustainable development and survival of DRT systems. The present study has a few limitations. First, the results of this study describe the associations between the built environment and DRT travel behavior. An association itself is not enough to establish an underlying causal effect of a built environment characteristic on DRT travel behavior (Schutt 2004; Singleton and Straits 2005). Furthermore, we were not allowed to collect data on all of the individual or household socioeconomic attributes of CB subscribers due to privacy concerns. Their effects on CB travel behavior cannot be disentangled from the constant terms. (As a result, the constant terms of the models are highly significant.)

Nevertheless, this study expands the literature on the associations between built environment attributes and DRT travel behavior. First, multiple dimensions of built environment characteristics have significant impacts on DRT use. Areas with greater residential populations, fewer employment opportunities, more residential land uses, less administrative land uses, poor connectivity to road networks and parking supply, better accessibility to different facilities, and long distances from urban business centers are conducive to the development of CB services (routes). Therefore, the origin stop of the home-to-work DRT routes should be located in urban areas with these built environment characteristics to absorb regular demands fully. Furthermore, giving more attention to long-distance trips can help ensure the competitiveness of CB services with respect to its alternatives in service quality and may enhance the superiority of DRT in commuting trips. Additionally, the differences in built environment effects between CB travel behavior and regular transit travel behavior indicate that maintaining a complementary role of DRT services to regular bus services will help absorb steady passenger flows. Tapping into niche markets that are not served by regular transit services or are served with poor service quality is a potential marketing direction for DRT systems. The findings of this study provide not only a theoretical basis for the identification of demand and potential markets for CB systems in China but also theoretical references for DRT systems worldwide that are designed for commuters.

References

Abdullah, Z. 2017. Customised bus services set to go on trial in 2018 [WWW Document]. *The straits times*. Available from https://www.straitstimes.com/singapore/transport/customised-bus-services-set-to-go-on-trial-in-2018. Accessed on 12.26.20.

Alonso-gonzález, M.J., O. Cats, N. Van Oort, S. Hoogendoorn-lanser, S. Hoogendoorn. 2021. What are the determinants of the willingness to share rides in pooled on—Demand services ? *Transportation* 48 (4): 1733–1765. Springer US. https://doi.org/10.1007/s11116-020-10110-2.

Anspacher, D., J. Khattak, A., Youngbin, Y. 2004. Demand-responsive transit shuttles: Who will use them? California PATH Working Paper. Berkeley.

Balcombe, R., R. Mackett, N. Paulley, J. Preston, J. Shires, H. Titheridge, M. Wardman, and P. White. 2004. *The demand for public transport: A practical guide, TRL limited*. Workingham, United Kingdom.

Cao, X. (Jason), S.L. Handy, and P.L. Mokhtarian. 2006. The influences of the built environment and residential self-selection on pedestrian behavior: Evidence from Austin. *TX. Transportation (Amst)* 33, 1–20. https://doi.org/10.1007/s11116-005-7027-2.

Cao, X. (Jason), P.L. Mokhtarian, S.L. Handy. 2009. Examining the impacts of residential self-selection on travel behaviour: A focus on empirical findings. *Transport Reviews* 29 (3): 359–95. https://doi.org/10.1080/01441640802539195.

Cao, X. (Jason). 2010. Exploring causal effects of neighborhood type on walking behavior using stratification on the propensity score. *Environment and Planning A* 42: 487–504. https://doi.org/10.1068/a4269.

Carmeron, A.C., and P.K. Trivedi. 1986. Econometric models based on count data: Comparisons and application of some estimators. *Journal of Applied Econometrics* 46: 29–53.

Cervero, R., and K. Kockelman. 1997. Travel demand and the 3Ds: Density, diversity, and design. *Transportation Research Part d: Transport and Environment* 2: 199–219. https://doi.org/10.1016/S1361-9209(97)00009-6.

Chen, X. (Michael), M. Zahiri, S. Zhang. 2017. Understanding ridesplitting behavior of on-demand ride services: An ensemble learning approach. *Transportation Research Part C: Emerging Technologies* 76: 51–70. https://doi.org/10.1016/j.trc.2016.12.018.

Chi, C.G.Q. 2012. An Examination of Destination Loyalty: Differences between First-Time and Repeat Visitors. *Journal of Hospitality and Tourism Research* 36: 3–24. https://doi.org/10.1177/1096348010382235.

Chu, F. 2015. Customized bus: Cherish what you have at the moment. *Transport Business China* 8: 52–55.

Currie, G., and N. Fournier. 2020. Why most DRT/Micro-Transits fail—What the survivors tell us about progress. *Research in Transportation Economics* 83: 100895. https://doi.org/10.1016/j.retrec.2020.100895.

Daniel, C.V. 2017. Beyond the bus: 'Microtransit' helps cities expand transportation services [WWW Document]. Available at https://www.governing.com/gov-transit-agencies-extend-service-with-microtransit.html. Accessed 1.18.21.

Davison, L., M. Enoch, T. Ryley, M. Quddus, and C. Wang. 2012. Identifying potential market niches for demand responsive transport. *Research in Transportation Business and Management* 3: 50–61. https://doi.org/10.1016/j.rtbm.2012.04.007.

Department Statista Research. 2016. Car ownership rate in China in 2016, by gender [WWW Document]. Available at https://www.statista.com/statistics/862200/china-car-ownership-rate-by-gender/. Accessed 5.11.20.

Ding, C., X. Cao, and Y. Wang. 2018b. Synergistic effects of the built environment and commuting programs on commute mode choice. *Transportation Research Part a: Policy and Practice* 118: 104–118. https://doi.org/10.1016/j.tra.2018.08.041.

Ding, C., Y. Lin, and C. Liu. 2014. Exploring the influence of built environment on tour-based commuter mode choice: A cross-classified multilevel modeling approach. *Transportation Research Part d: Transport and Environment* 32: 230–238. https://doi.org/10.1016/j.trd.2014.08.001.

Enoch, M., S. Potter, G. Parkhurst, M. Smith. 2004. INTERMODE: Innovations in demand responsive transport, final report. Department for Transport and Greater Machester Passenger Transport Executive, London.

Enoch, M., S. Potter, G. Parkhurst, M. Smith. 2006. Why do demand responsive transport systems fail? in *Transportation Research Board 85th Annual Meeting*. 22–26. https://doi.org/10.5860/choice.51-2973.

Ewing, R., and R. Cervero. 2010. Travel and the built environment. *Journal of the American Planning Association* 76: 265–294. https://doi.org/10.1080/01944361003766766.

Frank, L.D., M.A. Andresen, and T.L. Schmid. 2004. Obesity relationships with community design, physical activity, and time spent in cars. *American Journal of Preventive Medicine* 27: 87–96. https://doi.org/10.1016/j.amepre.2004.04.011.

Frank, L., M. Bradley, S. Kavage, J. Chapman, and T.K. Lawton. 2008. Urban form, Travel Time, and Cost Relationships with Tour Complexity and Mode Choice. *Transportation (Amst)* 35: 37–54. https://doi.org/10.1007/s11116-007-9136-6.

Guan, X., and D. Wang. 2019a. Influences of the built environment on travel: A household-based perspective. *Transportation Research Part a: Policy and Practice* 130: 710–724. https://doi.org/10.1016/j.tra.2019.10.003.

Guo, Z. 2013. Home parking convenience, household car usage, and implications to residential parking policies. *Transp Policy (Oxf)* 29: 97–106. https://doi.org/10.1016/j.tranpol.2013.04.005.

Handy, S. 2018. Enough with the "D's" already—Let's get back to "A" [WWW Document]. Transfers. URL https://transfersmagazine.org/enough-with-the-ds-already-lets-get-back-to-a/.

Huang, S.-W., and W.-J. Tsai. 2013. The analysis of measurements and influence factors of mixed land use. *Int J Biosci Biochem Bioinforma* 3: 206–210. https://doi.org/10.7763/ijbbb.2013.v3.197.

Huo, Y., Z. Qiu, Y. Xin, S. Dai, R. Pan, and Y. Wen. 2021. Exploring the influencing factors on demand-responsive customized bus ridership using multi-source data. In *The 6th International Conference on Transportation Information and Safety*, 293–299. Wuhan, China.

Imaz, A., K.M. Nurul Habib, A. Shalaby, and A.O. Idris. 2014. Investigating the factors affecting transit user loyalty. *Public Transport* 7: 39–60. https://doi.org/10.1007/s12469-014-0088-x.

Jain, S., N. Ronald, R. Thompson, and S. Winter. 2017. Predicting susceptibility to use demand responsive transport using demographic and trip characteristics of the population. *Travel Behaviour and Society* 6: 44–56. https://doi.org/10.1016/j.tbs.2016.06.001.

Jarrett, W. 2019. What is "Microtransit" for? [WWW Document]. *Human Transit*. Available at: https://humantransit.org/2019/08/what-is-microtransit-for.html. Accessed 1.18.21.

Kirby, R.F., and K.U. Bhatt. 1975. An analysis of subscription bus experience. *Traffic Quarterly* 29: 403–425.

Kirby, R.F., K.U. Bhatt. 1974. *Guidelines on the operation of subscription bus services.* Washington, DC.

Kitamura, R., and P.L. Mokhtarian. 1997. A micro-analysis of land use and travel in five neighborhoods in the San Impacts of travel-based multitasking on mode choice and value of time View project Travel satisfaction View project. *Transportation (Amst)* 24: 125–158. https://doi.org/10.1023/A:1017959825565.

Lavieri, P.S., and C.R. Bhat. 2019. Modeling individuals' willingness to share trips with strangers in an autonomous vehicle future. *Transportation Research Part a: Policy and Practice* 124: 242–261. https://doi.org/10.1016/j.tra.2019.03.009.

Laws, R. 2009. *Institutional Tepository Evaluating publicly-funded DRT schemes in England and Wales.* Loughborough University.

Li, Y. 2019. First AI bus debuts in Shanghai with customized services [WWW Document]. Available at http://www.ecns.cn/news/2019-09-18/detail-ifznxzmn1176640.shtml. Accessed 1.24.21.

Liu, T., and A. Ceder. 2015a. Analysis of a new public-transport-service concept: Customized bus in China. *Transp Policy (Oxf)* 39: 63–76. https://doi.org/10.1016/j.tranpol.2015.02.004.

Liu, T., A.A. Cedera, R. Bologna, and B. Cabantous. 2016b. Commuting by customized bus: A comparative analysis with private car and conventional public transport in two cities. *Journal of Public Transportation* 19: 55–74. https://doi.org/10.5038/2375-0901.19.2.4.

Ma, X. 2015. Why passengers do not accompany with customized bus? *Decision-Making* 10: 82–83.

Ma, J., Y. Yang, W. Guan, F. Wang, T. Liu, W. Tu, and C. Song. 2017. Large-scale demand driven design of a customized bus network: A methodological framework and beijing case study. *Journal of Advanced Transportation.* https://doi.org/10.1155/2017/3865701.

Mageean, J., and J.D. Nelson. 2003. The evaluation of demand responsive transport services in Europe. *Journal of Transport Geography* 11: 255–270. https://doi.org/10.1016/S0966-6923(03)00026-7.

Morin, D.A. 1974. Commuter bus clubs are showing the "professionals" how to meet transit demand. *Traffic Engineering, Inst Traffic Engr* 44: 24–27.

Munshi, T. 2016. Built environment and mode choice relationship for commute travel in the city of Rajkot, India. *Transportation Research Part d: Transport and Environment* 44: 239–253. https://doi.org/10.1016/j.trd.2015.12.005.

Ndibatya, I., and M. Booysen. 2020. Transforming paratransit in Africa ' s congested cities : An ICT—Enabled Integrated Demand Responsive Transport (iDRT) approach. In *IST-Africa 2020 Conference Proceedings*, eds. M. Cunningham, P. Cunningham, 1–10, Kampala, Uganda.

Ogra, A., and R. Ndebele. 2013. The role of 6Ds : Density, diversity, design, destination, distance, and demand management in transit oriented development (TOD). *Neo-International Conference on Habitable Environments 539–546.*

Palmer, K., M. Dessouky, and Z. Zhou. 2005. Factors influencing productivity and operating cost of demand responsive transit. *Transportation Research Part A: Policy and Practice.* 42 (3): 503–523. https://doi.org/10.1016/j.tra.2007.12.003.

Potts, J.F., M.A. Marshall, E.C. Crockett. 2010. A guide for planning and operating flexible public transportation services. https://doi.org/10.17226/22943.

Sabouri, S., K. Park, A. Smith, G. Tian, and R. Ewing. 2020. Exploring the influence of built environment on Uber demand. *Transportation Research Part D* 81: 102296. https://doi.org/10.1016/j.trd.2020.102296.

Schasché, S.E., and R.G. Sposato. 2021. *Systematic literature review of demand-responsive transport services.* Carinthia.

Schutt, R.K. 2004. *Investigating the social world: The process and practice of research*, 4th ed. Thousand Oaks, CA: Pine Forge Press.

Scott, R.A. 2010. *Demand responsive passenger transport in low-demand situations.* New Zealand.

Sehatzadeh, B., R.B. Noland, and M.D. Weiner. 2011. Walking frequency, cars, dogs, and the built environment. *Transp Res Part A Policy Pract* 45: 741–754. https://doi.org/10.1016/j.tra.2011.06.001.

Singleton, R., and B. Straits. 2005. *Approaches to social research*, 4th ed. New York: Oxford University Press.

StataCorp,. 2017. *Multilevel mixed-effects negative binomial regression*. College Station, TX: Stata Press.

Sultana, Z., S. Mishra, C.R. Cherry, M.M. Golias, and S. Tabrizizadeh Jeffers. 2018. Modeling frequency of rural demand response transit trips. *Transp Res Part A Policy Pract* 118: 494–505. https://doi.org/10.1016/j.tra.2018.10.006.

Wang, C., M. Quddus, M. Enoch, T. Ryley, and L. Davison. 2014. Multilevel modelling of Demand Responsive Transport (DRT) trips in greater Manchester based on area-wide socio-economic data. *Transportation (Amst)* 41: 589–610. https://doi.org/10.1007/s11116-013-9506-1.

Wang, C., M. Quddus, M. Enoch, T. Ryley, and L. Davison. 2015. Exploring the propensity to travel by demand responsive transport in the rural area of Lincolnshire in England. *Case Stud Transp Policy* 3: 129–136. https://doi.org/10.1016/j.cstp.2014.12.006.

Wang, J., T. Yamamoto, and K. Liu. 2019a. Role of customized bus services in the transportation system: Insight from actual performance. *Journal of Advanced Transportation* 2019: 6171532. https://doi.org/10.1155/2019/6171532.

Wang, J., T. Yamamoto, and K. Liu. 2020b. Key determinants and heterogeneous frailties in passenger loyalty toward customized buses: An empirical investigation of the subscription termination hazard of users. *Transportation Research Part C: Emerging Technologies* 115: 102636. https://doi.org/10.1016/j.trc.2020.102636.

Wang, J., T. Yamamoto, and K. Liu. 2021. Spatial dependence and spillover effects in customized bus demand: Empirical evidence using spatial dynamic panel models. *Transp Policy (Oxf)* 105: 166–180. https://doi.org/10.1016/j.tranpol.2021.03.004.

Weinberger, R. 2012. Death by a thousand curb-cuts: Evidence on the effect of minimum parking requirements on the choice to drive. *Transp Policy (Oxf)* 20: 93–102. https://doi.org/10.1016/j.tranpol.2011.08.002.

Yan, N. 2014. A customized bus market survey record [WWW Document]. Available at http://www.chinabuses.com/buses/2014/0729/article_14514.html. Accessed 3.31.18.

Yang, L., Y. Ao, J. Ke, Y. Lu, and Y. Liang. 2021. To walk or not to walk? Examining non-linear effects of streetscape greenery on walking propensity of older adults. *Journal of Transport Geography* 94: 103099. https://doi.org/10.1016/j.jtrangeo.2021.103099.

Yang, W., and X. Cao. 2018. Examining the effects of the neighborhood built environment on CO_2 emissions from different residential trip purposes: A case study in Guangzhou, China. *Cities* 81: 24–34. https://doi.org/10.1016/j.cities.2018.03.009.

Yang, S., Y. Fan, W. Deng, and L. Cheng. 2019. Do built environment effects on travel behavior differ between household members? A case study of Nanjing, China. *Transp Policy (Oxf)* 81, 360–370. https://doi.org/10.1016/j.tranpol.2017.12.006.

Zhan, H.L., and J.S. Dong. 2014. Qingdao launched the first customized bus in China. *Transp. Bus. China* 9: 20–23.

Zhang, W., D. Lu, Y. Chen, and C. Liu. 2021. Land use densification revisited: Nonlinear mediation relationships with car ownership and use. *Transp Res D Transp Environ* 98: 102985. https://doi.org/10.1016/j.trd.2021.102985.

Zhang, J, D.Z.W. Wang, and M. Meng. 2017. Analyzing customized bus service on a multimodal travel corridor: An analytical modeling approach. *Journal of Transportation Engineering Part A Systems* 143: 04017057. https://doi.org/10.1061/jtepbs.0000087.

Zhang, W., Y. Zhao, X. (Jason) Cao, D. Lu, and Y. Chai. 2020. Nonlinear effect of accessibility on car ownership in Beijing: Pedestrian-scale neighborhood planning. Transportation Research Part D: Transport and Environment 86: 102445. https://doi.org/10.1016/j.trd.2020.102445.

Zhao, D., G.P. Ong, W. Wang, and X.J. Hu. 2019. Effect of built environment on shared bicycle reallocation: A case study on Nanjing, China. *Transp Res Part A Policy Pract* 128: 73–88. https://doi.org/10.1016/j.tra.2019.07.018.

Chapter 9
Non-Linear and Threshold Effects of Built Environment on Customized Bus Markets

Abstract The high failure rate of demand-responsive transit (DRT) systems suggests that DRT services are viable only in selected areas. However, few studies have quantitatively examined how built environment characteristics affect DRT use. Applying gradient boosting decision trees to the data of customized bus service (CB, a type of DRT) in Dalian, we investigate the nonlinear association between the built environment and CB use, controlling for demographics and service features. Local accessibility at the residence and workplace is the most important correlates of CB use, followed by the proximity of workplace to bus stop. Some built environment variables (including distance from workplace to transit stops, distance from residence to business centers, and population density) influence CB use differently from traditional transit use observed in the literature. Furthermore, built environment variables show salient threshold associations with CB use, guiding planners to design the CB system efficiently.

Keywords Machine learning · Demand-responsive transit · Built environment · Travel behavior · Gradient boosting

9.1 Introduction

During the last decade, CB has been intensively promoted and implemented in some medium and large Asian cities because of its high quality, flexibility, and convenience (Abdullah 2017; Liu and Ceder 2015). By the end of 2021, 29 provinces in China had launched CB routes, with an annual ridership of nearly 180 million (Xinhuanet 2021). CB, primarily oriented towards commuters, is a type of demand-responsive transit (DRT) system. DRT is a demand-driven and user-oriented transportation service system that aims to provide a convenient, flexible, and personalized transit service to various destinations, such as workplaces, shopping areas, and medical facilities (Currie and Fournier 2020; Davison et al. 2012; Jain et al. 2017; Kirby and Bhatt 1974; Morin 1974). It accommodates on-demand travel requests and sometimes completely replaces traditional bus services in low-demand areas. As a supplement

J. Wang et al., *Customized Bus: User Travel Behavior and Demand Evolution*, Urban Sustainability, https://doi.org/10.1007/978-981-96-3488-0_9

to the traditional transit system, DRT improves individuals' mobility, enhances the quality and efficiency of their travel, and helps address the issue of social exclusion (Jain et al. 2017; König and Grippenkoven 2020; Laws et al. 2008; Mageean and Nelson 2003).

Despite the benefits, many DRT systems have failed in practice (Currie and Fournier 2020; Davison et al. 2014). A systematic investigation of about 120 DRT systems worldwide showed that 40% of the systems were out of business within three years of operation and 50% lasted less than seven years (Currie and Fournier 2020). A key reason for the high failure rate is a lack of prior knowledge on which areas are appropriate for DRT service (Davison et al. 2012). DRT systems are not designed for the entire market but for specific service areas. Outside of the areas, it is difficult for them to maintain a profitable or sustainable operation. Some qualitative studies have suggested that poor traditional transit systems, dispersed land-use patterns, areas with a complex (e.g., overly circuitous) street network, and underdeveloped regions (especially geographically isolated areas) are not conducive to the survival and development of DRT systems (Davison et al. 2012; Enoch et al. 2004; Mageean and Nelson 2003; Travers 1990; Wang et al. 2021). However, few quantitative studies have examined the associations between built environment characteristics and DRT use. Without empirical corroborations, the relationships are still speculative.

CB is usually designed to serve routine trips, particularly commuting trips (for example, the service in Luxembourg (Ducoli 2018), Pau, Madrid, Lyon (Padam Mobility 2020), and many Chinese cities). Therefore, it is different from traditional DRT systems. Although CB has more regular demand than DRT in general, it also serves a niche market. Therefore, it is a challenge to sustain CB in the long term (Chu 2015; Ma 2015; Yan 2014). Given the potential influence of the built environment on DRT development, a better understanding of built environment correlates of CB demand could be crucial to its sustainable operation.

In this study, we employ the gradient boosting decision trees (GBDT) approach to data from Dalian, China, to explore the association between the built environment and CB use. We emphasize the following research questions: (1) Which built environment characteristics of residence and workplace have important impacts on CB use? (2) Do built environment characteristics have nonlinear or threshold associations with CB use? A nonlinear association means that the effect of a variable varies by the value of the variable. A threshold effect is a special nonlinear effect, where the effect changes drastically once the variable exceeds a particular value (Galster 2018).

To the best of our knowledge, this is the first study that emphasizes the nonlinear association of the built environment with DRT demand in general and CB demand in particular. The answers to the research questions will inform system planners and operators of key built environment correlates of CB demand and their threshold ranges that are the most conducive to CB ridership. They fill the gap in the literature. Furthermore, the results show that the relationships between several built environment variables and CB use differ from their associations with traditional transit. This may fundamentally change planners' understanding of the connection between the built environment and CB demand.

The remainder of this paper is structured as follows. Section 9.2 provides background to the creation of CB and its design/operation details. Section 9.3 reviews previous studies regarding how the built environment affects DRT system operation exemplifying research gaps. Section 9.4 describes the data, variables, and modeling approach. Section 9.5 presents the results. Section 6 recapitulates the key findings and discusses associated policy implications.

9.2 Literature Review

Because CB has recently emerged, few studies have addressed the correlates of CB users' travel behavior and the mechanisms underlying the correlations. To shed light on the potential impacts of built environment characteristics on CB demand, we extend the scope of our review to DRT systems in general and summarize the important factors of DRT use. Previous studies suggest that three types of factors may affect DRT use and ridership (Sultana et al. 2018; Wang et al. 2015):

- service-/trip-related factors, such as travel time (Chen et al. 2017; Nelson and Phonphitakchai 2012; Wang et al. 2020), costs (Davison et al. 2014; Wang et al. 2021), quality of service (Davison et al. 2012), trip purpose (Lavieri and Bhat 2019), and trip distance (Wang et al. 2021);
- socio-economic and demographic characteristics, such as gender (Jain et al. 2017), age (Jain et al. 2017; Wang et al. 2015), and private vehicle ownership (Wang et al. 2014); and
- the characteristics of service areas, particularly built environment characteristics (Laws 2009; Wang et al. 2015, 2014, 2021).

A vast body of literature has demonstrated that the built environment, indicating spatial characteristics of a space, has a decisive influence on shaping people's travel behavior (Cao et al. 2009; Ewing and Cervero 2010). Although planners have defined the built environment in various ways, most studies have followed the measurement of "5Ds", namely, *density, diversity, design, distance to transit, and destination accessibility* (Deng 2022; Ewing and Cervero 2010; Yang et al. 2019). Accordingly, this current study adopted the same built environment dimensions.

Unlike traditional bus services with fixed routes and timetables, DRT systems are demand-driven and user-oriented, and require higher operating costs (Currie and Fournier 2020). They are designed to collect passengers and serve special travel needs with carpooling services and cannot be universally adapted to all human settlements. Their development and long-term survival are largely influenced by built environment characteristics of the areas they serve (Enoch et al. 2006, 2004; Laws 2009). Therefore, ascertaining the impacts of built environment characteristics on DRT use is essential to identifying the markets of potential DRT systems, developing effective operational strategies, and ensuring their long-term operation.

Most studies on built environment characteristics of DRT systems rely on a comparative analysis of their actual operational statistics and offer a qualitative assessment of the characteristics that are appropriate for DRT development. Enoch et al. (2006) investigated 72 DRT projects worldwide and found that highly dispersed land uses and partially developed areas are not conducive to DRT development. Enoch et al. (2004) explored factors contributing to the failure of DRT systems and identified that the underdeveloped areas with poor accessibility do not work well for DRT. A possible reason is that dispersed land uses and poor accessibility promote private vehicle ownership and use, which in turn are negatively associated with DRT ridership. For instance, Wang et al. (2014) found that private vehicle ownership has a negative effect on the demand of a DRT system in the metropolitan region of Greater Manchester. Furthermore, poor street connectivity is not conducive to the development of DRT systems. In particular, the widespread of cul-de-sacs and curvilinear street networks may induce longer travel distances and lower vehicle speeds, both of which contribute to longer travel times (Enoch et al. 2006). This issue is especially notable when picking up and dropping off passengers.

On the other hand, the areas with poor traditional transit services are important potential markets for DRT systems (Enoch et al. 2006, 2004; Jain et al. 2017). DRT is a supplement in the areas without convenient access to transit stops or high quality of transit service. Jain et al. (2017) reviewed the literature on DRT systems in Europe and America and identified poor access to train stations as a key parameter affecting the usability of DRT. Moreover, some DRT systems are designed to serve low-density areas where transit demand is low and traditional transit service is poor and inefficient, such as micro-transit systems operated in small cities, suburban areas, and rural areas (Enoch et al. 2004; Scott 2010; Volinski and Raton 2019). These areas often have low population and employment densities (Sultana et al. 2018; Volinski and Raton 2019).

A few quantitative studies have offered some evidence for built environment correlates of DRT use. Wang et al. (2015) applied an ordered logit model to investigate the correlates of DRT use in the rural area of Lincolnshire, England. They found that residents living in densely populated areas have a lower propensity to use DRT. Wang et al. (2014) developed a multilevel linear regression model to explore the correlates of DRT demand in Greater Manchester. They concluded that the influence of transit services is insignificant. Huo et al. (2021) explored the correlation between DRT ridership and land use types (measured as points of interest). They found that scenic spots and governmental agencies are positively associated with ridership, whereas companies and medical facilities show negative correlations with ridership. Besides the limited effort, a common issue of these studies is that they were not designed to emphasize the built environment. They considered a few built environment dimensions and overlooked the potential role of other dimensions.

Furthermore, few studies have considered the built environment at the destination (Wang et al. 2015, 2014). A transit trip usually requires first- and last-mile connections. Conceptually, the built environment at both ends of transit trips should influence individuals' transit use. Empirically, previous studies have demonstrated that built environment characteristics around destinations have a significant influence on the choice of conventional commuting modes (Ding et al. 2018b; Liu et al. 2021; Sun et al. 2017). Likewise, built environment characteristics around both residences and workplaces could influence DRT use. Ignoring built environment variables at workplaces may lead to omitted variable bias (Ding et al. 2018b).

Moreover, few studies on the correlates of DRT use (Wang et al. 2015, 2014) have assumed a linear or generalized linear association between them. Recently, an increasing number of empirical studies have applied machine learning models to examine the nonlinear correlations between built environment characteristics and travel behavior. They found threshold effects and identified the ranges of built environment variables within which they affect travel behavior most effectively (Ding et al. 2018a, 2018b; Ghaffar et al. 2020; Yang et al. 2021). For example, using a random forest algorithm, Chen et al. (2021) demonstrated that most built environment variables exhibited nonlinear impacts on transit use. They reported that the use was substantially higher in the communities where the entropy index of land use mix exceeds 0.5. Ding et al. (2019) also concluded that the effects of built environment characteristics on station-level ridership vary by the level of these characteristics, and that the effect of a variable may reach a plateau when it exceeds a certain value. From the perspective of system operation and planning, ascertaining these thresholds or intervals helps transit operators develop operational strategies efficiently (Ding et al. 2019; van Wee and Handy 2016).

9.3 Method

9.3.1 Data and Variables

The data used in this study came from the Dalian CB system. Dalian, located in Northeast China, is one of the major coastal cities in China. It plays a leading role in the economic development of Liaoning Province. The population of Dalian is about 7.45 million according to the 7th National Population Census. Besides traditional bus transit, the city is served by five subway lines, with 500,000 rail riders per day.

The Dalian CB system was established in early 2014. With more than 300 meticulously designed routes, this is one of the most successful systems in China. It serves more than 4,000 passengers per day. Because the system is designed for commuting, the demand is relatively regular, i.e., between residences and workplaces on weekdays. After its normal operation in December 2015, on average, about 8% of the subscriptions are from new users each month and the remaining passengers have subscribed to the CB before. Because of the small share of new users, existing users' subscription frequency largely determines the ridership of a CB route. The lower the frequency is, the smaller the ridership is, and the more difficult it is for the route to survive. In other words, a high turnover rate is detrimental to the survival of CB routes. Thus, this study emphasizes user subscription frequency from a planning perspective and identifies its built environment correlates, controlling for other factors.

In this study, we used passengers' monthly subscription records from October 2016 to September 2017. A monthly subscription allows a passenger to use the CB in the month when the subscription is ordered. We focused on the passengers who had started their subscription before the study period to ensure that all the individuals were familiar with the CB and had the opportunity to access the service. After data cleaning and filtering, our sample included 2,718 passengers. In total, these passengers subscripted to the CB 18,453 times during the one-year study period. Figure 9.1 and Fig. 9.2 illustrate the locations of their residences and workplaces. The residences were scattered throughout the whole region. By contrast, most of their workplaces were concentrated in a few economic development zones.

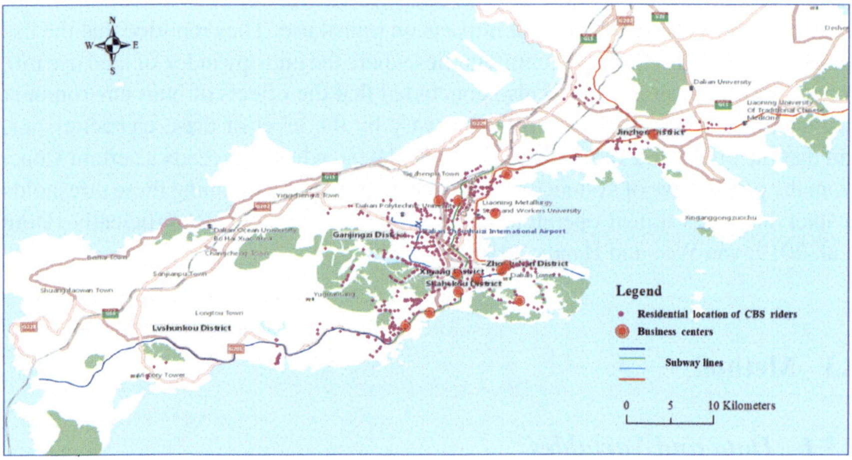

Fig. 9.1 Residential location of CB riders

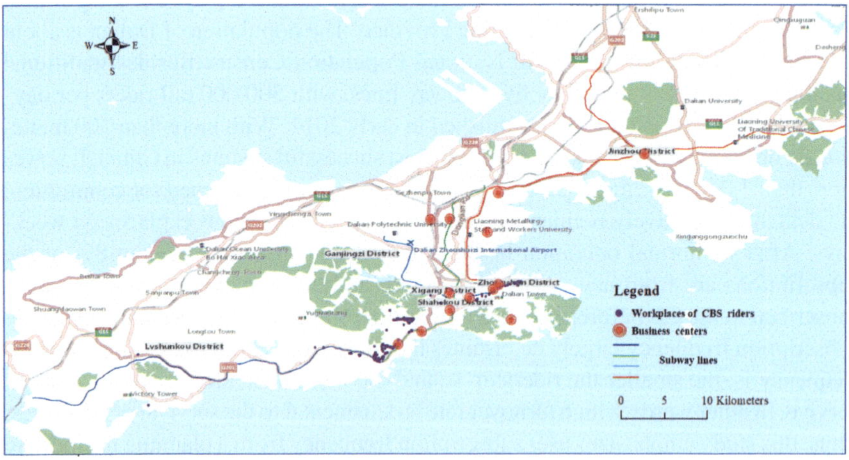

Fig. 9.2 Workplaces of CB riders

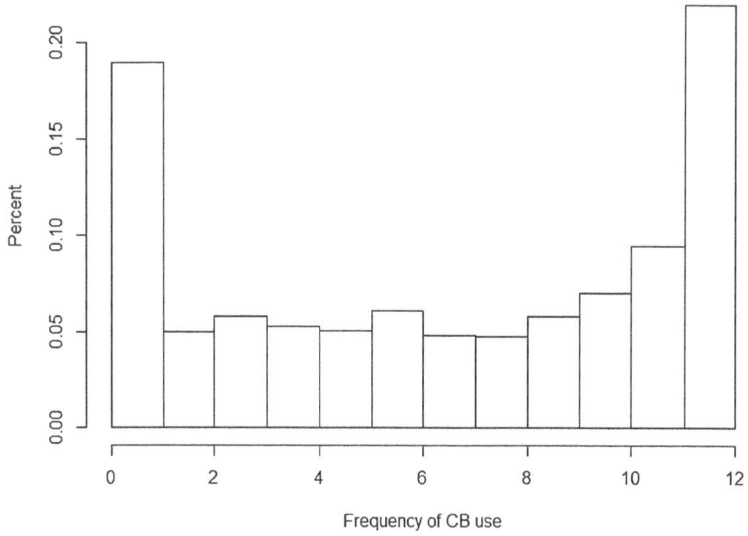

Fig. 9.3 Distribution of the CB subscription frequency

Figure 9.3 illustrates the frequency distribution of CB passengers. Passenger subscription frequency, a measure of CB use, ranges from 0 to 12, where "0" means that an individual has subscribed to the CB system before but did not purchase the service during the study period and "12" means that a passenger has kept riding the CB during the period.

One practical barrier to investigating built environment effects on travel behavior in China is the absence of detailed data on the built environment. To overcome this, we collected, mined, and integrated data from as many sources as we could obtain, including:

- 2015 urban land use data from the Bureau of Natural Resources, Dalian. We classified eight types of land uses, including administrative and public services, commercial and business facilities, green spaces and parks, industrial and manufacturing uses, residential areas, transportation, municipal utilities, and logistics and warehouses ((MOHURD) Ministry of Housing and Urban–Rural Development of the People's Republic of China 2012);
- population and employment information based on cellular signal data;
- point of interest (POI) data from the Gaode Map (including all types of POIs at the end of 2018); and
- vector geographic data from the Open Street Map.

Using the data above, we generated 5Ds of the built environment, including two density measures (population density and employment density), three measures of land use mix (average distance to the eight types of land uses, entropy index, and dissimilarity index), a measure of road network design (street connectivity), distance to transit stops, and distance to business centers. Except for the density measures, the variables were measured both at users' residences and workplaces. Because some built environment variables are highly correlated, we manually dropped several unimportant ones. In particular, the correlation coefficient between employment density at workplace and distance from workplace to land uses is –0.7, suggesting that multicollinearity may be at work. We dropped employment density because the model showed that it was much less important in predicting CB use than the distance variable. Similarly, we removed entropy indices and dissimilarity indices from the model because they are highly correlated with distances to land uses. Table 9.1 presents the final list of independent variables.

Data sources Gender and service attributes were derived from passenger subscription data. Housing transaction data were from Fang Holdings Limited (www.fang.com). Population data were derived based on cellular signal data. Land use data were obtained from Bureau of Natural Resources, Dalian. Road data were from the Open Street Map. Transit data and business center locations were based on the data of points of interest achieved by the Gaode Map. GDP data were calculated based on the statistical GDP data in 2010.

Service attributes included subscription price, travel time saving, and demand variation. All of them were computed based on the subscription records. Regarding individual characteristics, we included gender, housing price, and GDP. Because of the privacy requirement, passengers' income data are not available. Alternatively, we used GDP and housing price as proxies for passenger income. Using the National Census, we computed the GDP in the area within 1 km of passengers' residence. We also extracted the sales price data of second-hand houses in Dalian from Fang Holdings Limited (www.fang.com) and calculated the average housing price in the area.

9.3.2 Analysis Approach

In this study, we employed a GBDT model (Friedman 2001) to explore the association between the built environment and CB use. The model can help us explore the potential relationships between input features and the response variable, while automatically accounting for complex interactions among these input features (Molnar 2021).

The objective of the GBDT algorithm is to minimize a loss function $L(y_i, \rho)$. Its application can be divided into several steps. In Step 1, the approximation function $\hat{f}(X)$ of the response variable y is initialized as a constant:

$$\hat{f}(X_i) = arg\min_{\rho} \sum_{i=1}^{N} L(y_i, \rho) \tag{9.1}$$

where X is a vector of input features; ρ is the step length; i is the index of observations, and $i \in \{1,2, \ldots, N\}$. Because the response variable investigated in this study is a count variable, a Poisson distributed loss function is applied.

Table 9.1 Description and statistics of independent variables

Variables	Description	Mean	S.D	Min	Max
Individual characteristics					
Female	A dummy variable, which is equal to 1 for females and 0 for males	0.76	0.43	0	1
GDP	Total GDP within the 1000 m buffer of residence (based on the statistical GDP data in 2010. 10^4 RMB)	7.77	3.31	0.01	12.84
Housing price	Average housing price in the area around residence (housing sales price per m^2 in Jul. 2016. 10^3 RMB)	9.89	1.45	5.17	13.40
Service attributes					
Price per kilometer	Fare per kilometer (RMB/km)	0.65	0.22	0.22	1.50
Travel time saving	Travel time by CB – travel time by traditional transit service (min)	26.60	14.22	25.00	60.42
Demand variation	Standard deviation of the density of CB users within the 1000 m buffer of residence (based on the subscription data from Aug. 2015 to Sept. 2016. person/km^2/h)	1.55	1.87	0.00	6.84
Built environment characteristics					
Population density at residence	The number of people (in 2018) within the 1000 m buffer of residence (10^3 persons/3.14 km^2)	14.15	8.19	0.73	34.97
Distance from residence to land uses	Mean distance from residence to the nearest establishment for 8 types of land use (km)	1.31	0.88	0.04	4.10
Distance from workplace to land uses	Mean distance from workplace to the nearest establishment for 8 types of land use (km)	1.19	0.53	0.60	4.07
Connectivity at residence	The number of intersections within the 1,000 m buffer of residence	37.92	51.22	0	398

(continued)

Table 9.1 (continued)

Variables	Description	Mean	S.D	Min	Max
Connectivity at workplace	The number of intersections within the 1,000 m buffer of workplace	31.73	44.46	0	361
Distance from residence to rail station	Distance from residence to the nearest rail transit station (km)	1.48	1.23	<0.01	11.95
Distance from workplace to rail station	Distance from workplace to the nearest rail transit station (km)	1.26	0.48	<0.01	3.00
Distance from residence to bus stop	Distance from residence to the nearest bus stop (km)	0.22	0.31	<0.01	0.99
Distance from workplace to bus stop	Distance from workplace to the nearest bus stop (km)	0.26	0.15	<0.01	0.64
Distance from residence to business center	Distance from residence to the nearest business center (km)	5.11	3.16	<0.01	32.88
Distance from workplace to business center	Distance from workplace to the nearest business center (km)	3.07	0.75	<0.01	14.61

Note We also measured employment density at workplace, land use entropy index and dissimilarity index at residence and workplace. They were dropped out of the model because they were highly correlated with distances to land uses

In Step 2, we calculate the negative gradient (i.e., the residual of the previous estimate) for the *m*-th iteration (where m = 1 to M) as follows:

$$r_{im} = -\left[\frac{\partial L(y_i, f(X_i))}{\partial f(X_i)}\right]_{\hat{f}_m(X_i)=\hat{f}_{m-1}(X_i)} \tag{9.2}$$

In Step 3, we use a regression tree $g(X) = E(r|X)$ to fit r_{im} based on features X_i. Suppose that the regression tree has K terminal nodes, the optimal step length of each node can be calculated by:

$$\rho_k = \underset{\rho}{argmin} \sum\nolimits_{X_i \epsilon R_k} L\left(y_i, \hat{f}(X_i) + \rho\right) \tag{9.3}$$

where R_k is samples used to define the terminal node k. It can be subsampled from the training dataset or sampled from the whole training dataset. Then, $\hat{f}(X)$ is updated as:

$$\hat{f}(X) = \hat{f}(X) + \gamma \rho_{k(X)} \tag{9.4}$$

where $k(X)$ indicates the terminal node that an observation with features X would fall into; γ is the shrinkage or learning rate, with $0 < \gamma \leq 1$.

Finally, we repeat the above process for M times. The optimal model largely relies on a collective choice of hyper-parameters, including shrinkage, number of trees, number of splits for each tree, etc. The shrinkage value can be adjusted to scale the contribution of each tree to prevent the over-fitting problem (Friedman 2001). The number of trees and the number of splits for each tree also need to be specified and adjusted to capture the complex interactions between input features (Greenwell 2021; Ridgeway 2007).

When developing the model, we employed a general grid search method to tune the hyper-parameters (Boehmke and Greenwell 2019). The final model used the following hyper-parameters[1]: the learning rate was 0.005; the maximum tree depth was 4; the minimum number of observations in each terminal node was 11; the number of trees was 325. Note that we used a tenfold cross-validation procedure throughout the hyper-parameter optimization process to minimize overfitting and obtain a robust model. The Efron's Pseudo R-square of the final model is 0.116.

9.4 Results

9.4.1 Relative Importance of Independent Variables

Table 9.2 presents the relative importance of independent variables (i.e., input features) in predicting CB use (i.e., rider subscription frequency). The relative importance of an independent variable is measured by its contribution to reducing the squared loss, as compared to other independent variables. The larger the relative importance, the stronger the predictive power.

Among the variables in the model, distance from residence to land uses is the most important predictor and it contributes about a quarter of the predictive power. Distance from workplace to land uses, the second most important one, contributes about 10%. Because distance to land uses measures the average proximity of a geographical location to different land uses and is commonly used as an indicator of local accessibility (Zhang et al. 2020, 2021) and land use mix, these findings highlight the effect of accessibility to diverse land uses on CB use. Ranking third, distance from workplace to bus stop contributes 9.5%.

[1] For the grid search, we examined the learning rate ranging from 0.001 to 0.01 with an increment of 0.001, the number of trees ranging from 50 to 1000 with an increment of 25, the maximum tree depth ranging from 2 to 10 with an increment of 1, and the minimum number of observations for each node ranging from 5 to 20 with an increment of 1.

Table 9.2 Relative importance of independent variables

Variable	Rank	Relative importance (%) [a]	Total (%)
Individual characteristics			6.8
Female	14	2.0	
GDP	12	3.0	
Housing price	16	1.8	
Service attributes			16.1
Price per kilometer	11	3.2	
Travel time saving	9	4.4	
Demand variation	4	8.5	
Built environment characteristics			77.1
Population density at residence	10	4.4	
Distance from residence to land uses[b]	1	24.7	
Distance from workplace to land uses	2	9.9	
Connectivity at residence	13	2.7	
Connectivity at workplace	15	1.9	
Distance from residence to rail station	8	4.5	
Distance from workplace to rail station	7	4.6	
Distance from residence to bus stop	6	5.5	
Distance from workplace to bus stop	3	9.5	
Distance from residence to business center	5	7.9	
Distance from workplace to business center	17	1.5	

[a]The relative importance of all independent variables adds up to 100%
[b]Land uses refer to eight types of land uses, including administrative and public services, commercial and business facilities, green spaces and parks, industrial and manufacturing uses, residential areas, transportation, municipal utilities, and logistics and warehouses

9.4.2 Associations Between CB Use and Independent Variables

In this subsection, we will illustrate the accumulated local effects (ALE) plots of the most important independent variables in the model and elaborate on their relationships with CB use. We will also briefly discuss the remaining variables. However, please refer to the Appendix for their ALE plots. It is worth noting that ALE plots are centered at zero and the value along the vertical axis of an ALE plot represents the main effect of an independent variable at a certain value on the dependent variable compared to the average effect. We use the ticks on the horizontal axis to illustrate the distribution of the independent variable.

9.4.2.1 Built Environment Attributes

Distance from residence to land uses and distance from workplace to land uses contribute the most to predicting rider subscription frequency among all the built environment variables in the model. As illustrated in Fig. 9.4, both variables have a negative association with CB use. This finding appears consistent with the literature because a negative relationship was also found in some American and European DRT systems (Enoch et al. 2004). To begin with, a longer average distance to land uses suggests poor accessibility, which is conducive to private vehicle travel but discourages the use of traditional transit and similar modes (Ewing and Cervero 2010; Guan and Wang 2019). CB mostly complements traditional bus services and competes with private vehicles (Liu and Ceder 2015; Wang et al. 2019). Moreover, the longer the average distance to different types of land uses is, the more dispersed the land-use patterns of a neighborhood are. Dispersed land-use patterns impose challenges on the CB route and stop design (Enoch et al. 2006, 2004). They may result in long detours while passengers are picked up and dropped off, which increases the travel time of CB and reduces its reliability.

The ALE plots presented in Fig. 9.4 also illustrate the threshold effects of distances to land uses at residence and workplace on CB use. Passengers subscribe to CB for commuting most frequently when the average distance from residence to land uses is less than 1.8 km and the average distance from workplace to land uses is less than 1 km. The threshold at residence is larger than that of at workplace. This is plausible because passengers have more options to connect with CB stops at residence than those at workplace, where walking is the primary mode. Therefore, they are more sensitive to accessibility at workplace. Furthermore, the effect of distance to land uses at residence decreases sharply after it reaches the threshold, while the effect of the distance measure at workplace reduces stepwise.

Figure 9.5 shows a positive effect of distance from workplace to transit stops on CB use. The farther passengers' workplace is from traditional transit stops, the more often they subscribe to CB. This finding demonstrates that CB complements

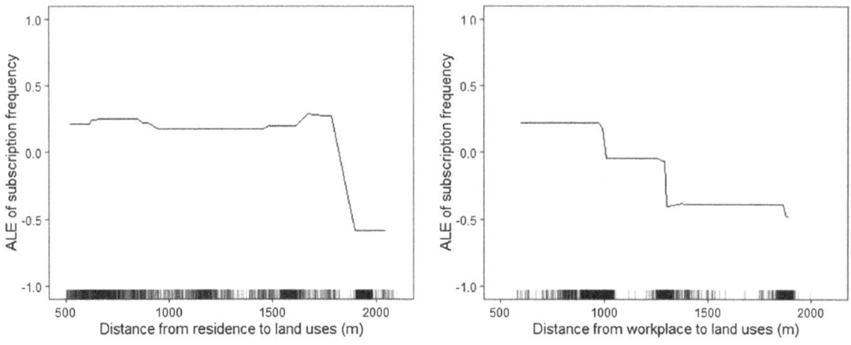

Fig. 9.4 The influences of distance to land uses on CB use

the traditional transit system in service coverage areas. This pattern makes sense because the areas that are poorly served by traditional transit are potential markets for DRT services (Currie and Fournier 2020; Enoch et al. 2004; Wang et al. 2021). Passengers' subscription frequency reaches plateaus when their workplace is about 400 m away from the nearest bus stop and 1,500 m away from the nearest subway station. These findings suggest that subway has a larger catchment area than bus transit, congruent with the common knowledge in the literature.

Passengers' subscription frequencies drop substantially when their workplace is less than 100 m from the nearest bus stop and less than 250 m from the nearest subway station. Furthermore, the size of the reduced effect for rail transit is about twice as large as that for bus transit. Both statistics imply that rail transit is more attractive in commuting mode choice than bus transit. More importantly, they suggest that employment opportunities that are very close to traditional transit stops are not conducive to CB development.

A comparison between the effects of distances to transit stops at workplace and those at residence suggests that the former is more critical to CB use than the latter. The relative importance of distances from residence to bus stop and rail station is similar to that of distance from workplace to rail station. However, the ALE plots in Fig. 9.6 show that the positive effect of distance from residence to bus stop on CB use is trivial and distance from residence to rail transit station has no clear pattern of influences. Although traditional transit is available around passengers' residential neighborhoods, its quality of service may be inferior to CB. For example, many neighborhoods where CB riders live are served by a limited number of bus routes and residents may have to make transfers to reach their workplaces (Wang et al. 2019); commuters may not be able to find a seat during transfer trips, especially during peak hours. By contrast, CB is a point-to-point service and guarantees a seat for each passenger (Liu and Ceder 2015; Wang et al. 2019). These differences may make CB more competitive than traditional transit.

Figure 9.7 illustrates that distance from residence to the closest business center has a positive relationship with CB use. This finding differs from the negative effect on traditional transit systems commonly observed in the literature (Cervero 2006;

a) Distance to bus stop

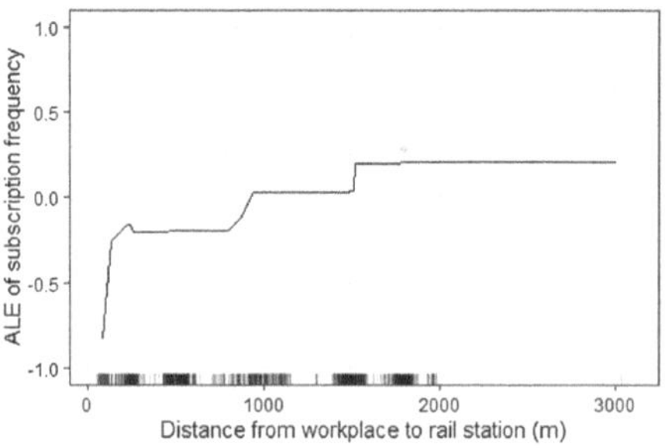

b) Distance to rail transit station

Fig. 9.5 The influences of distance from workplace to transit stops on CB use

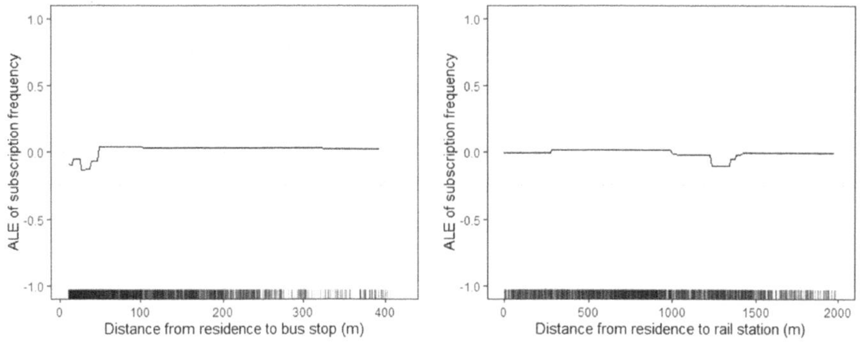

a) Distance to bus stop

b) Distance to rail transit station

Fig. 9.6 The influences of distance from residence to transit stops on CB use

Ding et al. 2014). Generally, employment opportunities are concentrated at business centers and their surrounding areas. The farther individuals live from the business center, the longer their commuting distance tends to be. The positive relationship may be observed because CB is more competitive for longer-distance trips (Wang et al. 2019, 2021; Zhang et al. 2017). The positive effect also suggests that CB can compete with private vehicles and its widespread application helps reduce car use. Moreover, its threshold effect suggests that residential communities within 3 km from business centers are less conducive to the long-term operation of CB systems.

Population density has a small but non-trivial impact on CB use (Figure A1). Passenger subscription frequency is the highest when the density is smaller than about 6,300 persons per square kilometer (=20,000/3.14). The negative effect of

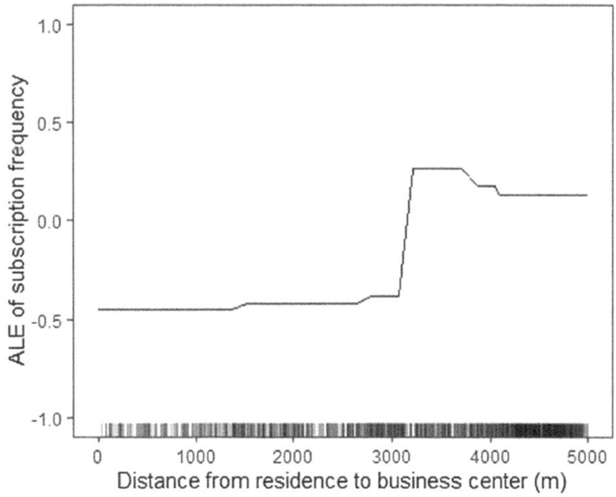

Fig. 9.7 The influence of distance from residence to business center on CB use

population density on CB use differs from its positive effect on the use of traditional transit (Cervero 2002; Ewing and Cervero 2010). High density and a critical mass are positively associated with the supply of traditional transit services. That is, residents in densely developed areas have plenty of transit options and are more likely to use traditional transit. However, the CB system is designed to complement traditional transit systems and its market is oriented towards lower-density areas. Distance from workplace to the closest business center and street connectivity at both residence and workplace have minuscule effects on CB use (Figures A2 and A3).

9.4.2.2 Other Attributes

The variation of CB demand contributes 8.5% of the predictive power, ranking fourth among all the variables in the model. However, its important role is often neglected in previous studies. Demand variation has a negative association with CB use (Fig. 9.8). The more volatile the demand is, the less frequently CB users subscribe to the service. CB operators sometimes need to adjust routes and stop locations based on residential and workplace locations of subscribed passengers. CB demand that is stable spatially and temporally may reduce the uncertainty of CB operation caused by the adjustments, lowering the burden on CB passengers and hence increasing its attractiveness.

Although it is believed that excessive demand variation is detrimental to DRT operation, existing studies have yet to analyze its effect on DRT use and potential thresholds. Some studies of best practices have recommended the ranges of demand density for DRT systems. For instance, Potts et al. (2010) stated that 5–11 passengers per hour are needed to sustain DRT operation efficiently. This current study shows

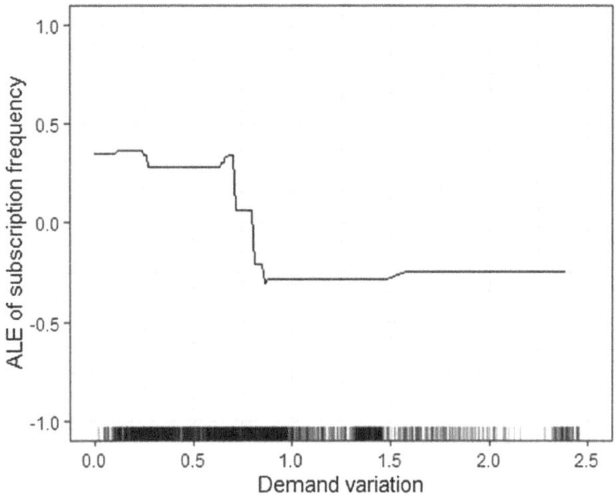

Fig. 9.8 The influence of demand variation on CB use

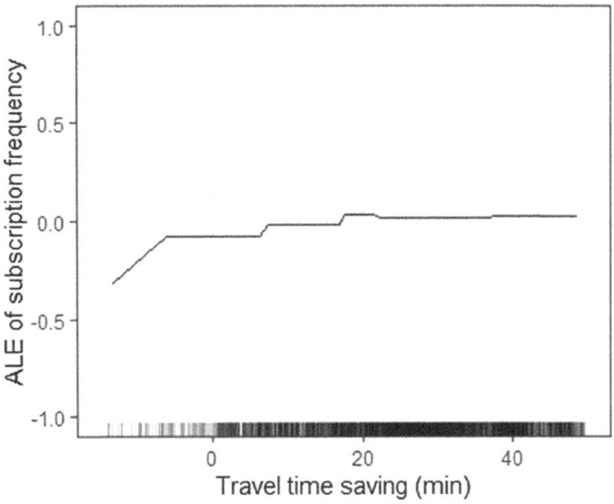

Fig. 9.9 The influence of travel time saving on CB use

that when the standard deviation of CB demand is smaller than 0.7 passengers per squared kilometer per hour, passengers have the highest subscription frequency.

Travel time saving has a positive association with CB use, consistent with the DRT literature (Yim and Ceder 2006). Interestingly, Fig. 9.9 shows that some riders chose CB even if its travel time was longer than that of traditional transit. This finding implies that direct travel, seat availability, and other benefits offered by CB outweigh travel time saving.

As we expected, fare has a negative association with CB use (Figure A4). The proxies for income (housing price and GDP) are positively associated with CB use (Figures A5 and A6). This relationship makes sense because the fare for CB is usually more than three times as much as traditional transit fare (Wang et al. 2019). Higher-income people have a higher willingness to pay for a premium service. Consistent with some previous studies (Nelson and Phonphitakchai 2012; Sultana et al. 2018), female users are more likely to subscribe to the CB more frequently (Figure A7).

9.5 Conclusions

This study explored the associations between built environment characteristics and the use of customized bus service, based on passengers' 12-month subscription data in Dalian. Using the gradient boosting decision trees approach to relax the linear assumption between variables, it illustrated the nonlinear and threshold relationships of built environment variables with CB use, controlling for individual characteristics and service features. This study has several limitations. First, the results present more of an association between built environment characteristics and CB use. The association itself is inadequate to infer the underlying causality (Schutt 2004; Singleton and Straits 2005). Future studies should conduct qualitative analysis to scrutinize why individuals choose to use CB, why they stop using it, and what role built environment characteristics play in the decisions. The information helps planners better understand the mechanisms underlying quantitative results. Second, this study used big data, which contained little information on passengers' demographic characteristics. Model prediction could be improved if it can incorporate a comprehensive set of individual-level characteristics, such as age, income, and private vehicle ownership. Alternatively, future studies could employ primary data (by surveying both CB riders and non-riders) to model built environment correlates of individuals' choice of CB. The results of this study could help develop hypotheses in future studies. Third, this study used 5Ds to measure the built environment, which facilitates the comparison of built environment effects observed in this study to the literature. However, these measures do not capture the quality of land uses and transportation infrastructure, which may also influence CB use. When data are available, future studies should test these variables. Finally, because the data were from a high-density city, some results may not be generalizable to low-density cities. Nevertheless, this study uncovered novel findings on the relationships between built environment characteristics and CB use and provided evidence-based implications for the planning and design of CB systems.

First, built environment variables predict CB use. More importantly, the associations of several variables with CB use differ from their relationships with traditional transit. In particular, distance from workplace to bus stop, distance from workplace to rail station, distance from residence to bus stop, and distance from residence to business center are positively associated with CB use; population density has a negative association with CB use. On the other hand, distance from workplace to land uses and distance from residence to land uses are negatively associated with CB use, congruent with the relationships with traditional transit. Other variables tested in the model have small or uncertain connections with CB use. Second, this study showed that built environment variables have salient nonlinear and threshold effects, largely ignored in the transit literature. In the remainder of this section, we will summarize the results of key built environment variables and discuss associated implications.

Local accessibility at both residence and workplace greatly impacts CB use, and higher accessibility facilitates passengers' subscription frequency. In particular, the residential neighborhoods that are on average less than 1.8 km away from different types of land uses and the workplaces that are on average less than 1 km away from different types of land uses are promising CB markets. This finding shed light on CB route design and stop location choice. CB operators often struggle with the decision of where to deploy terminal stops to attract an adequate number of passengers. The finding suggests that they should first analyze the accessibility of residential neighborhoods and workplaces to identify potential service areas for CB in the region. To identify riders efficiently in the first step of CB design, they should prioritize the areas for survey administration. Then, terminal stops should be deployed in the areas with the highest market potential and strategically extended to other priority areas.

This approach can also be used to identify additional market areas and potential stops along an existing CB route if it falls short of passenger demand. These new stops can help sustain the declining route.

Distance from workplace to transit stops is critical to CB use. We found that the workplaces that are close to traditional bus stops (less than 100 m) and subway stations (less than 250 m) are not conducive to CB use. Therefore, CB routes and stops at the workplace should avoid the areas with immediate access to traditional transit services. This finding corroborates that CB complements traditional transit in geographic coverage. Currently, CB operators in Dalian seldom coordinate service planning with traditional transit agencies. Their complementary nature calls for planning agencies to require coordinated planning and design of the two transit systems. The finding also informs CB operators of how the deployment of a new traditional transit route affects CB use. For instance, if a new metro station is deployed within 250 m from workplaces, it will be detrimental to the ridership of CB routes nearby. On the other hand, if it is more than 1,000 m from workplaces, it will have few impacts on CB ridership. If new bus stops are more than 200 m from workplaces, their impact on CB ridership is trivial. These thresholds provide CB operators with nuanced guidance for service adjustments.

Additionally, the residential communities that are less than 3 km from major business centers are associated with lower CB use. Very densely-developed residential neighborhoods are not conducive to CB use, either. Therefore, CB route planning should not emphasize these areas.

Appendix

See Figs. A1–A7

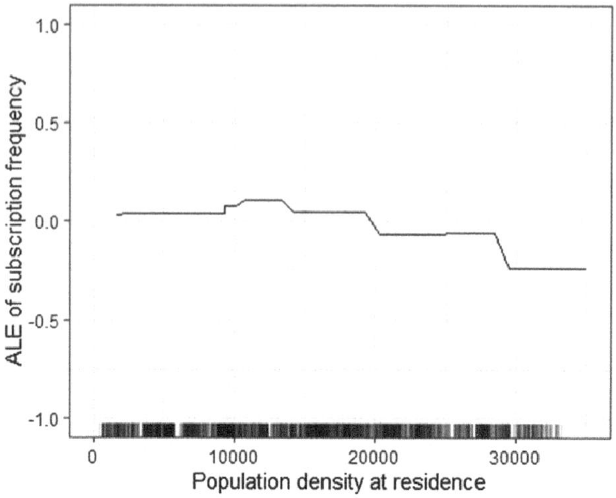

Fig. A1 The influence of population density at residence on CB use

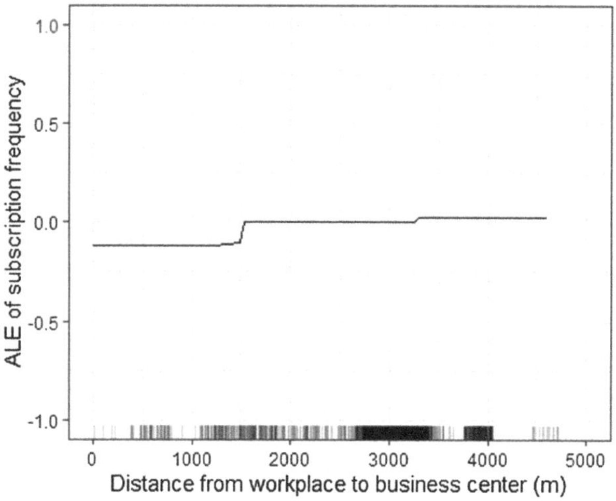

Fig. A2 The influence of distance from workplace to business center on CB use

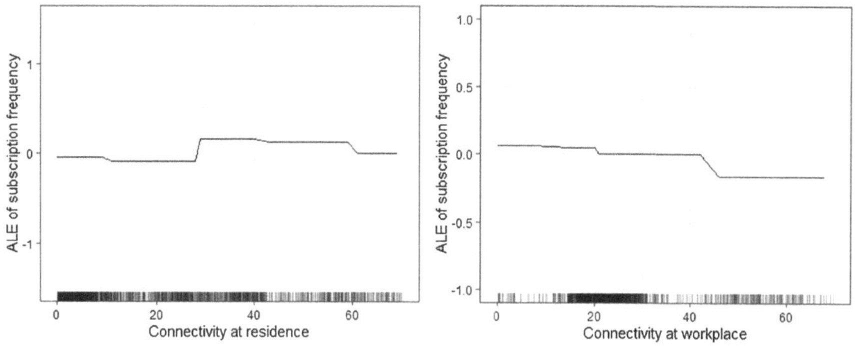

(a) Connectivity at residence (b) Connectivity at workplace

Fig. A3 The influences of street connectivity on CB use

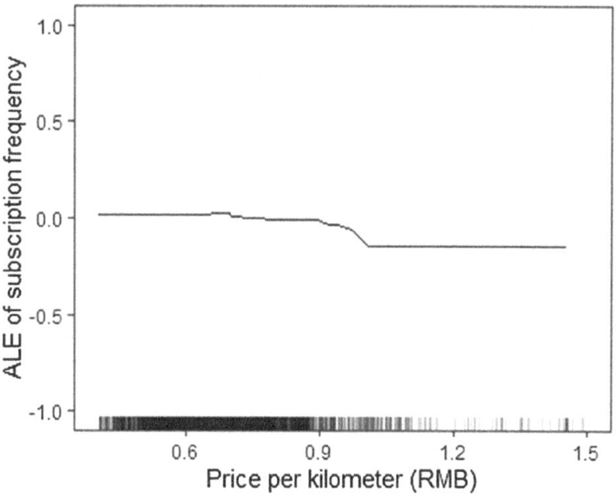

Fig. A4 The influence of subscription price on CB use

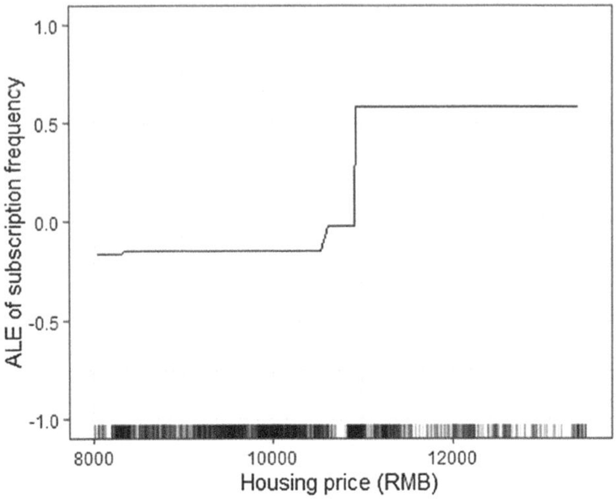

Fig. A5 The influence of housing price on CB use

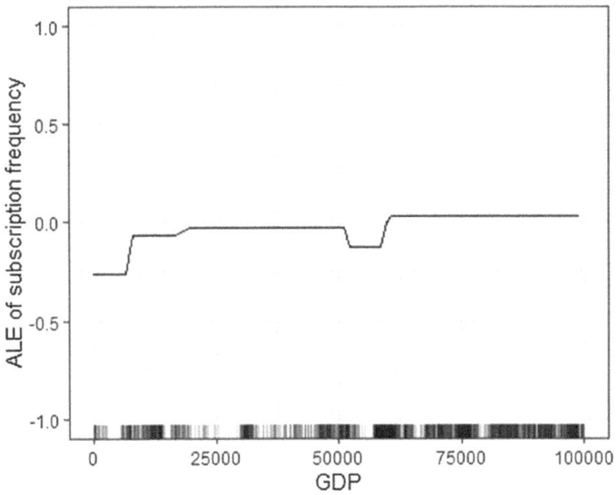

Fig. A6 The influence of GDP on CB use

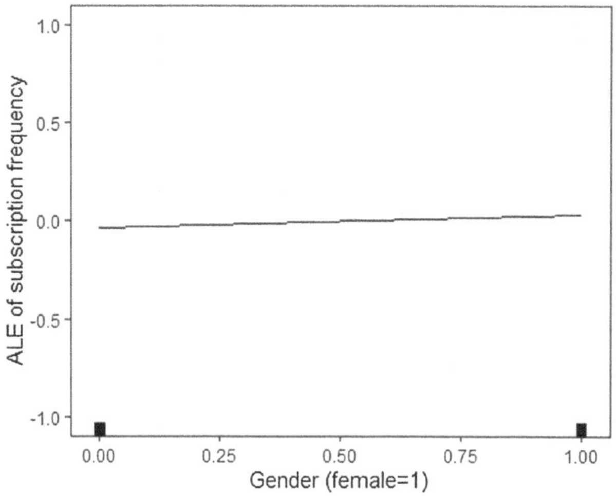

Fig. A7 The influence of gender on CB use

References

Abdullah, Z. 2017. Customised bus services set to go on trial in 2018 [WWW Document]. *The Straits Times*. https://www.straitstimes.com/singapore/transport/customised-bus-services-set-to-go-on-trial-in-2018 (accessed 12.26.20).

Boehmke, B., and B. Greenwell. 2019. *Hands-on machine learning with R*. Chapman and Hall/CRC.

Cao, X. (Jason), P.L. Mokhtarian, and S.L. Handy. 2009. Examining the impacts of residential self-selection on travel behaviour: A focus on empirical findings. *Transport Reviews*. https://doi.org/10.1080/01441640802539195

Cervero, R. 2002. Built environments and mode choice: Toward a normative framework. *Transportation Research Part d: Transport and Environment* 7: 265–284. https://doi.org/10.1016/S1361-9209(01)00024-4.

Cervero, R. 2006. Alternative approaches to modeling the travel-demand impacts of smart growth. *Journal of the American Planning Association* 72: 285–295. https://doi.org/10.1080/01944360608976751.

Chen, E., Z. Ye, and H. Wu. 2021. Nonlinear effects of built environment on intermodal transit trips considering spatial heterogeneity. *Transportation Research Part d: Transport and Environment* 90: 102677. https://doi.org/10.1016/j.trd.2020.102677.

Chen, X. (Michael), M. Zahiri, and S. Zhang. 2017. Understanding ridesplitting behavior of on-demand ride services: An ensemble learning approach. *Transportation Research Part D: Transport and Environment* 76: 51–70. https://doi.org/10.1016/j.trc.2016.12.018

Chu, F. 2015. Customized Bus: Cherish what you have at the moment. *Transport Business China* 8: 52–55.

Currie, G., and N. Fournier. 2020. Why most DRT/Micro-Transits fail—What the survivors tell us about progress. *Research in Transportation Economics* 83: 100895. https://doi.org/10.1016/j.retrec.2020.100895.

Davison, L., M. Enoch, T. Ryley, M. Quddus, and C. Wang. 2012. Identifying potential market niches for Demand Responsive Transport. *Research in Transportation Business and Management* 3: 50–61. https://doi.org/10.1016/j.rtbm.2012.04.007.

Davison, L., M. Enoch, T. Ryley, M. Quddus, and C. Wang. 2014. A survey of demand responsive transport in great Britain. *Transport Policy (Oxford)* 31: 47–54. https://doi.org/10.1016/j.tranpol.2013.11.004.

Deng, Y. 2022. Application of machine learning with a surrogate model to explore seniors' daily activity patterns. *Transportation Letters* 14, 972–982. https://doi.org/10.1080/19427867.2021.1969169.

Ding, C., Cao, X. (Jason), Næss, P., 2018a. Applying gradient boosting decision trees to examine non-linear effects of the built environment on driving distance in Oslo. *Transportation Research Part D: Transport and Environment* 110: 107–117. https://doi.org/10.1016/j.tra.2018.02.009

Ding, C., X. Cao, and C. Liu. 2019. How does the station-area built environment influence Metrorail ridership? Using gradient boosting decision trees to identify non-linear thresholds. *Journal of Transport Geography* 77: 70–78. https://doi.org/10.1016/j.jtrangeo.2019.04.011.

Ding, C., X. Cao, and Y. Wang. 2018b. Synergistic effects of the built environment and commuting programs on commute mode choice. *Transportation Research Part A: Policy and Practice* 118: 104–118. https://doi.org/10.1016/j.tra.2018.08.041.

Ding, C., Y. Lin, and C. Liu. 2014. Exploring the influence of built environment on tour-based commuter mode choice: A cross-classified multilevel modeling approach. *Transportation Research Part D: Transport and Environment* 32: 230–238. https://doi.org/10.1016/j.trd.2014.08.001.

Ducoli, L.L. 2018. Kussbus: Redefining the daily commute [WWW Document]. *Luxembourg Times*. https://www.luxtimes.lu/en/luxembourg/kussbus-redefining-the-daily-commute-602d5e03de135b92369040cf (accessed 10.28.22).

Enoch, M., S. Potter, G. Parkhurst, and M. Smith. 2004. *INTERMODE: Innovations in demand responsive transport, final report*. London: Department for Transport and Greater Machester Passenger Transport Executive.

Enoch, M., S. Potter, G. Parkhurst, and M. Smith. 2006. Why do demand responsive transport systems fail? In *Transportation Research Board 85th Annual Meeting*, 22–26. https://doi.org/10.5860/choice.51-2973

Ewing, R., and R. Cervero. 2010. Travel and the built environment. *Journal of the American Planning Association* 76: 265–294. https://doi.org/10.1080/01944361003766766.

Friedman, J.H. 2001. Greedy function approximation: A gradient boosting machine. *Annals of Statistics* 29: 1189–1232.

Galster, G.C. 2018. Nonlinear and threshold effects related to neighborhood: Implications for planning and policy. *Journal of Planning Literature* 33: 492–508. https://doi.org/10.1177/0885412218793693.

Ghaffar, A., S. Mitra, and M. Hyland. 2020. Modeling determinants of ridesourcing usage: A census tract-level analysis of Chicago. *Transportation Research Part C: Emerging Technologies* 119: 102769. https://doi.org/10.1016/j.trc.2020.102769.

Greenwell, B. 2021. Generalized boosted regression models [WWW Document]. https://www.rdocumentation.org/packages/gbm/versions/2.1.8/topics/gbm (accessed 2.6.21).

Guan, X., and D. Wang. 2019. Residential self-selection in the built environment-travel behavior connection: Whose self-selection? *Transportation Research Part D: Transport and Environment* 67: 16–32. https://doi.org/10.1016/j.trd.2018.10.015.

Huo, Y., Z. Qiu, Y. Xin, S. Dai, R. Pan, and Y. Wen. 2021. Exploring the influencing factors on demand-responsive customized bus ridership using multi-source data. In *The 6th International Conference on Transportation Information and Safety*, 293–299. Wuhan, China.

Jain, S., N. Ronald, R. Thompson, and S. Winter. 2017. Predicting susceptibility to use demand responsive transport using demographic and trip characteristics of the population. *Travel Behaviour and Society* 6: 44–56. https://doi.org/10.1016/j.tbs.2016.06.001.

Kirby, R.F., and K.U. Bhatt. 1974. *Guidelines on the operation of subscription bus services*. Washington, DC.

König, A., and J. Grippenkoven. 2020. The actual demand behind demand-responsive transport: Assessing behavioral intention to use DRT systems in two rural areas in Germany. *Case Studies on Transport Policy* 8: 954–962. https://doi.org/10.1016/j.cstp.2020.04.011.

Lavieri, P.S., and C.R. Bhat. 2019. Modeling individuals' willingness to share trips with strangers in an autonomous vehicle future. *Transportation Research Part A: Policy and Practice* 124: 242–261. https://doi.org/10.1016/j.tra.2019.03.009.

Laws, R. 2009. *Institutional tepository evaluating publicly-funded DRT schemes in England and Wales*. Loughborough University.

Laws, R., M. Enoch, S. Ison, and S. Potter. 2008. DRT schemes in England and Wales and considerations for their future. In: *87th Annual Meeting of the Transportation Research Board 2008*, 20. Washington, DC.

Liu, J., B. Wang, and L. Xiao. 2021. Non-linear associations between built environment and active travel for working and shopping: An extreme gradient boosting approach. *Journal of Transport Geography* 92: 103034. https://doi.org/10.1016/j.jtrangeo.2021.103034.

Liu, T., and A. Ceder. 2015. Analysis of a new public-transport-service concept: Customized bus in China. *Transport Policy (Oxford)* 39: 63–76. https://doi.org/10.1016/j.tranpol.2015.02.004.

Ma, X. 2015. Why passengers do not accompany with customized bus? *Decision-Making* 10: 82–83.

Mageean, J., and J.D. Nelson. 2003. The evaluation of demand responsive transport services in Europe. *Journal of Transport Geography* 11: 255–270. https://doi.org/10.1016/S0966-6923(03)00026-7.

Ministry of Housing and Urban-Rural Development of the People's Republic of China (MOHURD). 2012. Code for classification of urban land use and planning standards of development land.

Molnar, C. 2021. Interpretable machine learning: A guide for making black box models explainable [WWW Document]. https://leanpub.com/interpretable-machine-learning

Morin, D.A. 1974. Commuter bus clubs are showing the "professionals" how to meet transit demand. *Traffic Engineering—Institute of Transportation Engineers* 44: 24–27.

Nelson, J.D., and T. Phonphitakchai. 2012. An evaluation of the user characteristics of an open access DRT service. *Research in Transportation Economics* 34: 54–65. https://doi.org/10.1016/j.retrec.2011.12.008.

Padam Mobility. 2020. Boosting passenger experience for commuters with DRT [WWW Document]. Intelligent Transport. https://www.intelligenttransport.com/transport-whitepapers/106128/boosting-passenger-experience-for-commuters-with-drt/ (accessed 10.28.22).

Potts, J.F., M.A. Marshall, and E.C. Crockett. 2010. A guide for planning and operating flexible public transportation services. https://doi.org/10.17226/22943

Ridgeway, G. 2007. Generalized boosted models: A guide to the gbm package.

Schutt, R.K. 2004. *Investigating the social world: The process and practice of research*, 4th ed. Thousand Oaks, CA: Pine Forge Press.

Scott, R.A. 2010. Demand responsive passenger transport in low-demand situations. New Zealand.

Singleton, R., and B. Straits. 2005. *Approaches to social research*, 4th ed. New York. Oxford University Press.

Sultana, Z., S. Mishra, C.R. Cherry, M.M. Golias, and S. Tabrizizadeh Jeffers. 2018. Modeling frequency of rural demand response transit trips. *Transportation Research Part A: Policy and Practice* 118: 494–505. https://doi.org/10.1016/j.tra.2018.10.006.

Sun, B., A. Ermagun, and B. Dan. 2017. Built environmental impacts on commuting mode choice and distance: Evidence from Shanghai. *Transportation Research Part D: Transport and Environment* 52: 441–453. https://doi.org/10.1016/j.trd.2016.06.001.

Travers, M. 1990. Canterbury dial a bus study—Final report.

van Wee, B., and S. Handy. 2016. Key research themes on urban space, scale, and sustainable urban mobility. *International Journal of Sustainable Transportation* 10: 18–24. https://doi.org/10.1080/15568318.2013.820998.

Volinski, J., and B. Raton. 2019. Microtransit or general public demand-response transit services: State of the practice, CRP Synthesis of Transit Practice 141. The National Academies Press.

Wang, C., M. Quddus, M. Enoch, T. Ryley, and L. Davison. 2014. Multilevel modelling of Demand Responsive Transport (DRT) trips in Greater Manchester based on area-wide socio-economic data. *Transportation (Amst)* 41: 589–610. https://doi.org/10.1007/s11116-013-9506-1.

Wang, C., M. Quddus, M. Enoch, T. Ryley, and L. Davison. 2015. Exploring the propensity to travel by demand responsive transport in the rural area of Lincolnshire in England. *Case Studies on Transport Policy* 3: 129–136. https://doi.org/10.1016/j.cstp.2014.12.006.

Wang, J., T. Yamamoto, and K. Liu. 2019. Role of customized bus services in the transportation system: Insight from actual performance. *Journal of Advanced Transportation* 2019: 6171532. https://doi.org/10.1155/2019/6171532.

Wang, J., T. Yamamoto, and K. Liu. 2020. Key determinants and heterogeneous frailties in passenger loyalty toward customized buses: An empirical investigation of the subscription termination hazard of users. *Transportation Research Part C: Emerging Technologies* 115: 102636. https://doi.org/10.1016/j.trc.2020.102636.

Wang, J., T. Yamamoto, and K. Liu. 2021. Spatial dependence and spillover effects in customized bus demand: Empirical evidence using spatial dynamic panel models. *Transport Policy (Oxford)* 105: 166–180. https://doi.org/10.1016/j.tranpol.2021.03.004.

Xinhuanet. 2021. More than 29 provinces lanched customized bus services [WWW Document]. http://bj.news.cn/2021-11/14/c_1128062771.htm

Yan, N. 2014. A customized bus market survey record [WWW Document]. http://www.chinabuses.com/buses/2014/0729/article_14514.html (accessed 3.31.18).

Yang, L., Y. Ao, J. Ke, Y. Lu, and Y. Liang. 2021. To walk or not to walk? Examining non-linear effects of streetscape greenery on walking propensity of older adults. *Journal of Transport Geography* 94: 103099. https://doi.org/10.1016/j.jtrangeo.2021.103099.

Yang, S., Y. Fan, W. Deng, and L. Cheng. 2019. Do built environment effects on travel behavior differ between household members? A case study of Nanjing, China. *Transport Policy (Oxford)* 81: 360–370. https://doi.org/10.1016/j.tranpol.2017.12.006.

Yim, Y., and A. Ceder. 2006. Smart feeder/shuttle bus service: Consumer research and design. *Journal of Public Transportation* 9: 97–121. https://doi.org/10.5038/2375-0901.9.1.5.

Zhang, J., D.Z.W. Wang, and M. Meng. 2017. Analyzing customized bus service on a multimodal travel corridor: An analytical modeling approach. *Journal of Transportation Engineering, Part A: Systems* 143. https://doi.org/10.1061/jtepbs.0000087.

Zhang, W., D. Lu, Y. Chen, and C. Liu. 2021. Land use densification revisited: Nonlinear mediation relationships with car ownership and use. *Transportation Research Part D: Transport and Environment* 98: 102985. https://doi.org/10.1016/j.trd.2021.102985.

Zhang, W., Y. Zhao (Jason), X. Cao, D. Lu, and Y. Chai. 2020. Nonlinear effect of accessibility on car ownership in Beijing: Pedestrian-scale neighborhood planning. *Transportation Research Part D: Transport and Environment* 86: 102445. https://doi.org/10.1016/j.trd.2020.102445

Chapter 10
Spatial–Temporal Evolution of Customized Bus Demand

Abstract This paper proposes customized bus (CB) demand models to investigate dynamic adjustments, spatial dependence, and spatial spillover effects using spatial dynamic panel data techniques and a balanced panel data set collected over two years. The endogeneity introduced by spatial dependence effects is resolved by extending the moment restrictions of the system GMM estimator to a spatial autoregressive dynamic panel. A wide range of variables representing factors related to 1) characteristics of service supply, 2) demographic characteristics, and 3) land use and accessibility were mined and fused to investigate their direct and spillover effects on CB demand. The results reveal the mechanism of spatial dependence such that the increments (or decrement) in demand rather than the amount of demand itself in the neighborhood contribute to the spatial dependence effects. There is more than 43% matched demand generated in a given area for each-unit increment of demand from its neighborhood. CB services are more attractive for passengers with long-distance trips, and therefore, increasing the focus on long-distance trips is one potential strategy for increasing ridership. Determined by the role of CB in the transportation system, areas with poor accessibility are found to have larger market niches for CB services.

Keywords Spatial dependence · Spillover effects · Demand evolution · Customized bus · Spatial dynamic panel model

10.1 Introduction

As an innovative, demand-driven, and personalized type of bus transit service, customized bus (CB) services have emerged in many cities in China in recent years; these hold the promise of establishing more attractive and diverse public transportation (PT) services and thereby reducing urban traffic congestion during peak hours. However, lessons drawn from some failed CB operations cases drive us to attach

J. Wang et al., *Customized Bus: User Travel Behavior and Demand Evolution*, Urban Sustainability, https://doi.org/10.1007/978-981-96-3488-0_10

more importance to demand evolution, which has been proven to be a core ingredient of service design, operation, and even system survival but has not yet received sufficient attention (Wang et al. 2020).

Travel demand has been found to exhibit extensive spatial dependence theoretically and empirically, particularly in ridesharing systems (Becker et al. 2017; Friedrich and Noekel 2017; Ke et al. 2017; Ni et al. 2018). As with other ridesharing systems, a CB route potential occurs, if the temporal and spatial travel requests of a certain number of demanders match (Friedrich et al. 2018). More specifically, CB routing generally addresses time window constraints in picking up and dropping off passengers (Liu and Ceder 2015; Tong et al. 2017). With the expected boarding and arrival times of passengers matched, a CB can actively absorb passengers from areas it passes through, and therefore the spatial dependence of CB demand occurs. The spatial dependence effects are more significant in areas close to the origins and destinations of CB routes because the CBs are permitted to detour more in these areas to maximize ridership. However, travel demand may continue to exhibit the same trend as historical demand (Ke et al. 2017; Liu et al. 2019; Yao et al. 2018), a phenomenon that is generally regarded as time dependence. These characteristics pose challenges in CB demand modelling because of the need to address various spatial effects and temporal adjustments simultaneously.

Spatial dependence effects, spatial spillover effects, and temporal dependence effects are only barely accommodated simultaneously in the field of travel demand modelling and forecasting. Blending these spatiotemporal effects into a single function introduces endogeneity and increases the risk of overfitting (Elhorst 2012). As exemplified in Fig. 10.1, spatial dependence reflects an endogenous situation in which the CB demand observed for one unit, $y_{A,t}$, depends on the values of the neighbouring observations from nearby units, $y_{B,t}$ (LeSage and Pace 2009), while a spillover effect refers to an exogenous interaction in which the CB demand observed for a particular unit depends on certain independent explanatory variables of other units, $x_{B,t}$ (Elhorst 2014).

Regarding the estimation of demand for conventional public transportation systems, the related cross-sectional data and methodologies have the disadvantage that they do not allow for any adjustment of the period in the demand functions (Graham et al. 2009). An increasing number of studies are emphasizing the importance of temporal effects and have accordingly introduced univariate time-series approaches for demand modelling in conventional public transit systems. These approaches provide potential solutions for endogeneity and dynamics but suffer from the strong assumption that the patterns of historical demand will extend over time and, correspondingly, include no exogenous variables (Tsai et al. 2014). Hence, the univariate methods mentioned above ignore exogenous factors and individual heterogeneity, which should also contribute to the variation in demand.

With the increasing popularity of panel data methods, additional exogenous independent variables have been examined to reveal the effects of corresponding factors. Such methods also allow for the incorporation of unobserved individual-specific heterogeneity and thereby enable the control of the impact of omitted variables to reduce multi-collinearity (Hsiao 2005). A dynamic panel data model turns out to be

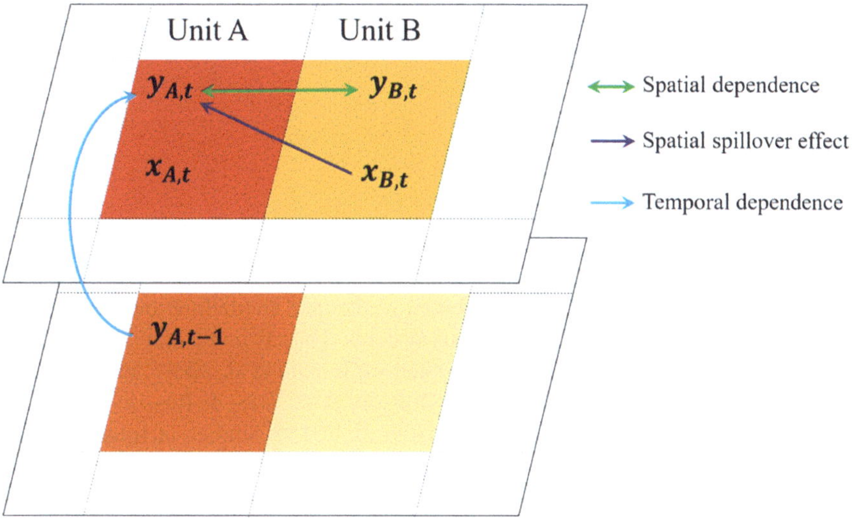

Fig. 10.1 The structure of spatiotemporal dependence

a superior solution for accommodating time-adjustment patterns to uncover dynamic relationships involving demand.

With the addition of spatial interactions, a panel data model can be extended to control spatial dependence and spillover effects. Observations from a particular location may significantly depend on values at other locations (Elhorst 2014). Additionally, changes in factors in a particular unit may impact the dependent variable not only in the same unit but also in surrounding areas (Elhorst 2017). Nevertheless, previous studies have typically paid more attention to temporal adjustments in public transit demand while ignoring the existence of spatial dependence and spillover effects, particularly for on-demand transit systems such as CB services.

In this study, we employed spatial dynamic panel data techniques to simultaneously accommodate dynamic adjustments, spatial dependence, spillover effects, and unit-specific heterogeneity. Variables representing a wide range of factors, including factors related to 1) characteristics of service supply, 2) land use and accessibility, and 3) demographic characteristics, were mined from operational data and additional geospatial big data sources and fused to investigate the effects on CB demand based on panel observations over 26 months.

This paper is structured as follows. Section 10.2 reviews previous studies relevant to demand modelling and forecasting methods and the primary factors incorporated into our models. Detailed descriptions of the study area and the continuous observations of CB demand are given in Sect. 10.3. Section 10.4 introduces the methods used to quantify the considered factors, which are drawn from a review of previous studies and may influence the assessment of CB demand, and the specifications of the general CB demand model. The results and discussion are presented in Sect. 10.5. Finally,

Sect. 10.6 summarizes the key findings of the study as drawn from the estimation results.

10.2 Literature Review

Accurate forecasts of demand are essential for efficient service planning, resource allocation, and operations in transportation systems. Over the past several decades, a great number of studies have contributed to demand modelling and forecasting for PT services in the context of local bus systems (Dargay and Hanly 2002), urban rail (Blainey and Preston 2010), metro systems (Graham et al. 2009), and entire PT systems on the urban and national scales (Boisjoly et al. 2018; de Rus 1990; FitzRoy and Smith 1998; Graham et al. 2009; Tsai et al. 2014). Studies of transit system demand can be classified into 1) studies based on traveler attitudes and perceptions and 2) studies based on associated environmental, system, and behavioral characteristics (Taylor et al. 2003). The primary approaches taken by studies in the first category are based on users' satisfaction with the transit services they have experienced. Generally, the relevant data are derived from questionnaire surveys and face-to-face interviews and therefore risk being highly subjective. Individual utility is the main concern in these studies. By contrast, studies in the second category can be regarded as causal analyses that are mainly based on univariate and/or multivariate models (Taylor et al. 2003). They have the advantage of being more sophisticated empirical studies that permit the employment of a wider range of data. To produce more robust and accurate results and inferences, a larger sample size is required, and a set of factors that is as comprehensive as possible should be examined using a wide variety of multivariate models.

Because of the types of data used, this study can also be regarded as a causal analysis of the demand for a special transportation system, in which demand models are developed to address the potential spatial dependence and spillover effects. There are few previous studies that have focused on these issues for on-demand transit systems. However, it is worth performing an extensive review of the literature on demand modelling for transportation systems, in order to gain insights into the appropriate selection of factors for this study. Based on secondary data sources and published/unpublished studies of urban surface transportation systems, (Paulley et al. 2006) undertook an analysis of the factors impacting the demand for PT in Great Britain to identify cost-effective schemes for improving services. Although a wide range of factors were examined, the influences of fares, quality of service, income, and car ownership were the most significant and thus treated as the focuses of the analysis. (García-Ferrer et al. 2006) estimated users' responses to changes in pricing and in the characteristics of PT services after controlling for nonstationary and strongly seasonal features of the data set through time-series models. (Bresson et al. 2003) analyzed the effects of fares, income, and service supply on the demand for PT using panel data from English counties and French urban areas and found that the demand was relatively sensitive to fare changes. The quality of service and characteristics of service supply in these studies were defined in terms of time-related attributes and other service-specific attributes, such as vehicle kilometers per capita (Balcombe et al. 2004; Bresson et al. 2003; Paulley et al. 2006).

In comparison to the factors related to the quality of service and characteristics of service supply, demographic characteristics are likely to have a less significant impact but may still be very important when interpreting discrepancies in travel patterns between areas (Balcombe et al. 2004; Pasha et al. 2016). Growth in population size has been demonstrated to statistically increase the demand for PT (Kain and Liu 1999; Taylor et al. 2003). The gross domestic product (GDP) per capita may impact PT demand both positively and negatively (Balcombe et al. 2004). The positive effect is attributed to the increase in travel resulting from an increase in GDP, while the negative effect is derived from the notable stimulus to own private vehicles.

Previous studies have also considered the effects of spatial characteristics on the demand for PT system usage. More specifically, land-use patterns, distance from the urban centre, settlement size, land-use mix, provision of local facilities, and accessibility of transportation networks have been used as indicators to measure certain aspects of spatial characteristics in demand models (Balcombe et al. 2004; Chakrabarti 2017; Chen and McKnight 2007; Pasha et al. 2016; Taylor et al. 2003; Tsai et al. 2014, 2012; Yang et al. 2019). These studies have demonstrated a strong overall relationship between variations in area type and demand for PT.

Previous studies have paid considerable attention to temporal adjustments using time-series and dynamic panel data models but have barely addressed the spatial dependence effects that arise in demand for PT, particularly in travel-sharing systems such as CB systems. (Ni et al. 2018) emphasized that urban travel flows exhibit extensive spatial autocorrelations caused by origin dependence, destination dependence, and/or origin–destination (O-D) dependence. These authors noted that travel flows might be spatially interdependent even though they do not share the same origin or destination, as long as they share neighboring origins and/or destinations. The effects of spatial dependence will be more significant in influencing the demand for travel-sharing systems. Through spatial regression, (Becker et al. 2017) investigated the effects of neighborhood features on demand for a free-floating car-sharing system and demonstrated the significance of spatial dependence and spillover effects.

The literature mentioned above provides an important foundation for CB demand modelling in terms of factor selection and methodology adoption. In summary, factors related to 1) the characteristics of CB service supply, 2) demographic characteristics, and 3) land use and accessibility should be considered in CB demand models. In addition, the spatial dependences that affect CB demand and spillover effects of the influencing factors will result in biased estimates and inferences if they are not properly addressed. To capture a clear picture of the demand evolution for a CB system, it is crucial to simultaneously accommodate dynamic adjustments, spatial dependence, and spatial spillover effects.

This study focuses on the impacts of endogenous interactions related to spatiotemporal dependence and the abovementioned factors. The interactions related to spatial dependence and dynamic adjustments can be derived from operational data and introduced into demand models with well-defined relationships based on spatial contiguity. Factors pertaining to the characteristics of service supply are also available from the operational data for a CB system. By contrast, factors representing information about land use, accessibility, and demographic characteristics are not provided by an observed demand data set. Instead, additional data sources, including national population data, economic census data, and geospatial big data, are needed to calculate and infer these attributes. Therefore, information from such additional data sources should be mined and fused with the operational data to construct ample explanatory variables.

10.3 Study Area and CB Demand

Dalian, one of the major coastal cities in Northeast China, is taken as the study area in this paper. Dalian contains more than 6.69 million people (National Bureau of Statistics of the People's Republic of China 2011) and 1.37 million private motor vehicles (Dalian Municipal Bureau of Statistics 2017) and is suffering from ever-increasing traffic congestion, especially during rush hours. However, the Dalian rail transit system is not well developed, and currently, only two subway lines and two light rail transit lines are in operation. The two subway lines primarily provide transport services within the main city area, while the two other light rail transit lines connect emerging city centres to the main city area. Therefore, Dalian's PT sector is eager to find a solution to the city's severe traffic congestion.

As a demand-driven, user-oriented, and customized transit system, a CB system treats travel request collection as a top priority. Prospective users are first required to register with the CB system and submit their travel requests, including the O-D information and the desired departure and arrival times (Liu and Ceder 2015). Based on this initially collected potential demand, transit operators design preliminary CB routes and release their origin areas, destination areas, and departure/arrival times online to recruit passengers. The operators choose and launch the CB routes for which sufficient passengers are recruited. The stops of the CB routes are placed at public places (such as the nearby conventional bus stations) and as close as possible to actual origins and destinations based on negotiations with passengers conducted via telephone. The boarding times will be sent to passengers through electronic messaging. The passengers are required to arrive at their boarding stations 2–3 min before the designed boarding times. A monthly subscription (that is valid for all workdays in a month) and a daily subscription (that is valid for a specific day) for the service are available.

Monthly subscriptions are held by the majority of those requesting service in the Dalian CB system, while the CB routes dominated by daily subscribers mainly provide service in a few popular travel corridors with relatively unpredictable subscriber numbers. Therefore, the evolution in demand for monthly subscriptions is more critical and is the focus of this paper. Dalian's CB system mainly provides services for commuters (Wang et al. 2019), transporting passengers from their homes to their workplaces in the morning and picking them up from their workplaces in the

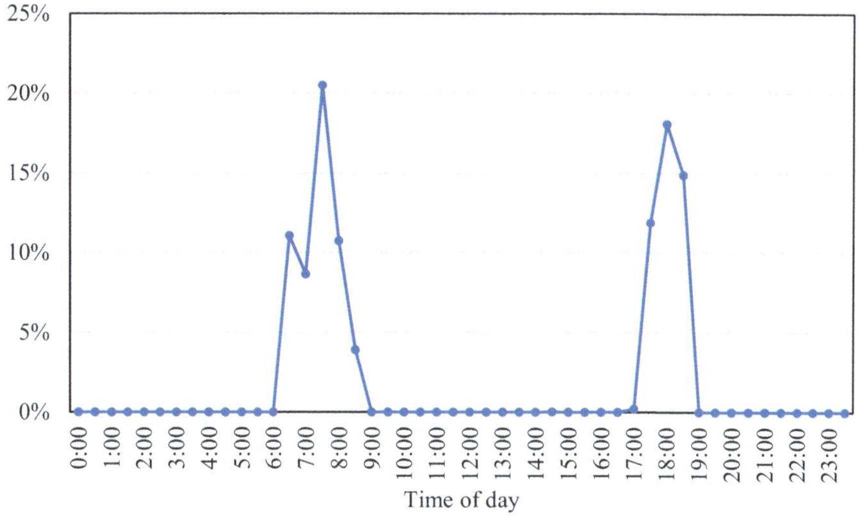

Fig. 10.2 Distribution of departure times for CB routes

afternoon. Thus, the CB trips (for monthly subscribers) occur in an intensive manner during the peak hours of Dalian's road network, i.e., residential pickups from 6:30 to 8:30 a.m. and workplace pickups from 5:00 to 6:30 p.m. (as illustrated in Fig. 10.2).

As a bus-ridesharing system, the CBs mainly provide services in areas with sufficient aggregated demand, i.e., the service areas covered by existing routes. The Dalian CB services were found to act as a complement to conventional PT services, particularly in areas with poor coverage or lacking conventional PT services (Wang et al. 2019), although the total coverage of conventional PT services is overwhelming. CB services are always designed to cater to specific travel requests, such as requests for customized commuter buses, customized school buses, and customized business buses (Liu and Ceder 2015), which determines that the number of passengers in the CB system cannot reach the levels of conventional PT and taxis.

The service level of CBs significantly outperforms that of conventional bus and rail transit services (Liu et al. 2016; Wang et al. 2019). Commuting by CB is more expensive than by conventional PT but is two-thirds cheaper than commuting by taxi (Wang et al. 2019). Therefore, CB services are viewed as an intermediate mode of transportation that represents a balance between travel cost and the quality of service typical of conventional PT services and taxis (Wang et al. 2019).

To model CB demand, the urban area of Dalian was uniformly partitioned into n 1 km $\times$ 1 km grid cells, each of which is regarded as a unit of analysis (panel). The demand in a unit is defined as the number of monthly subscriptions (orders) generated from that unit in one month. The actual CB demand for all these units was observed continuously from August 2015 to September 2017. The distribution of CB demand (home-to-work) for monthly subscriptions during the research period is shown in Fig. 10.3, while the spatial distributions of the origins and destinations of demand within the service areas are illustrated in Figs. 10.4 and 10.5.

Benefiting from the popularity gained from more than a year's trial operation (from April 2014 to March 2015), the daily subscription rapidly attracted a large

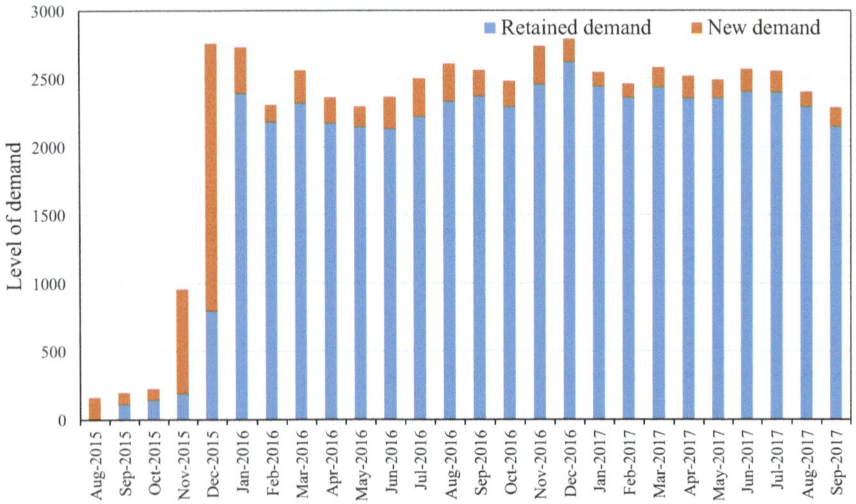

Fig. 10.3 Distribution of CB demand for monthly subscriptions

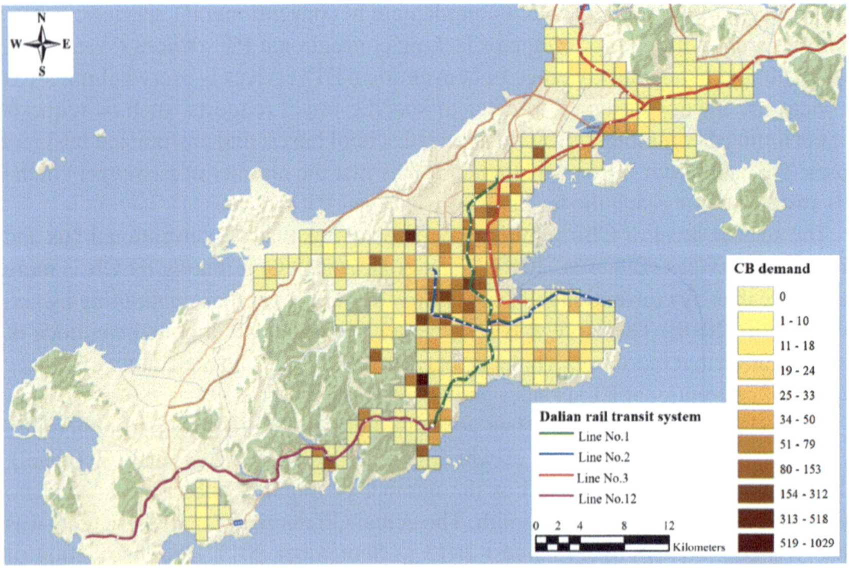

Fig. 10.4 Spatial distribution of origins of home-to-work CB demand (monthly subscriptions)

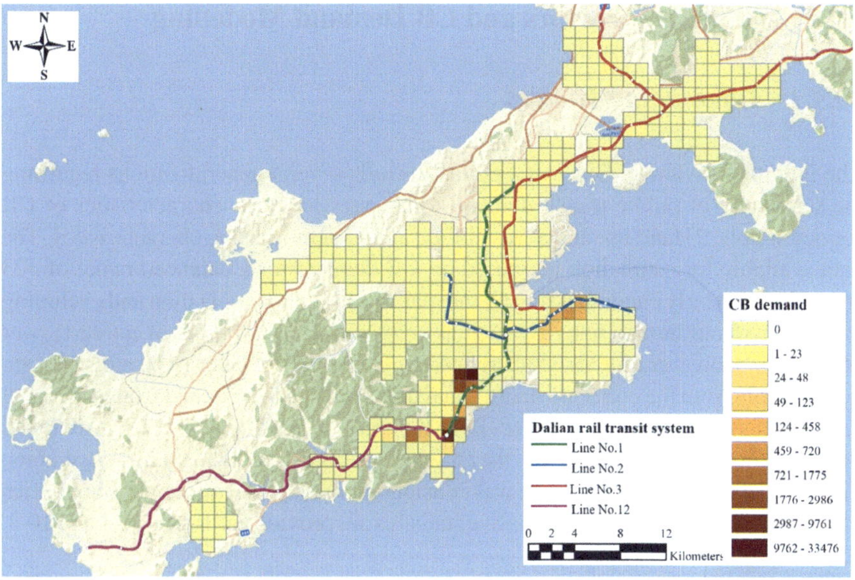

Fig. 10.5 Spatial distribution of destinations of home-to-work CB demand (monthly subscriptions)

number of users at the beginning of the official launch of the Dalian CB system, in June 2015, (Wang et al. 2019). Then, from August 2015, the monthly subscriptions were gradually dominating the CB system after months of growth. The demand evolution patterns between the "rapid growth period (pre-December 2015, as shown in Fig. 10.3)" and "steady operation period (post-December 2015)" may not be exactly the same. However, without the evidence of exogenous interventions (such as aggressive marketing strategies) observed, the entire research period (rather than the steady operation period only) is natural to be taken into account to explore the general evolution patterns of CB demand. The spatiotemporal evolution of the CB demand essentially determines the service operation (namely, the generation of new routes and adjustment of launching routes) during the entire progress of the CB system development.

Similar to the PT system, the CB service cannot cover the entire area of Dalian. As shown in Fig. 10.4, the areas covered with coloured cells are the service areas of the Dalian CB system. Owing to the absence of information about the exact boundary of the service areas, we included all the cells with demand generated and a 1000 m buffer (of demand) covered (but eliminated areas with very few people residents or few economic activities (POIs)) as the service area. The service areas cover almost all the built areas of Dalian city and are included in the demand model.

10.4 Influencing Factors and CB Demand Modelling

10.4.1 Influencing Factors

The literature review in Sect. 10.2 points towards several generalizations regarding the sensitivity of public transit demand to changes in 1) the characteristics of CB service supply, 2) land use and accessibility, and 3) demographic characteristics. The data available for estimation are based on CB demand for a balanced panel of 423 spatial units (nearly encompassing all the main built-up areas of Dalian and excluding the cells without human activity, such as cells without any points of interest) over 26 months. However, the variables related to factors in the latter two categories are not available from the collected data set and therefore had to be mined from additional data sources and properly quantified. Even though some cells are observed no demand generated during the study period, the entire service area (including 423 spatial units; Fig. 10.4) of the Dalian CB system was considered for unbiased inference. Summaries of variable descriptions and descriptive statistics are provided in Tables 10.1 and 10.2, respectively.

In terms of the characteristics of the CB service supply, the factors that determine the subscription price per kilometre, travel distance served, and route design for the Dalian CB service have been computed and considered in the CB demand modelling. In particular, with reserved seating, a comfortable in-vehicle environment, no transfers, and few midway stops, CB service has an advantage in the quality of service, particularly with respect to longer trips, over conventional PT (Wang et al. 2020; Zhang et al. 2017). This advantage can be observed because the long distance amplifies the passengers' perception of the difference in service quality. Additionally, route distance is typically employed as an important index of service supply in conventional PT systems (García-Ferrer et al. 2006). In this study, the shortest path distances for each passenger (between origin and destination) were fetched and aggregated by the unit as an indicator of CB service supply based on the digital vector map in the geographic information system.

Because the Dalian CB system employs a distance tariff scheme, the posted subscription price depends on travel distance and therefore cannot be directly included to investigate the price elasticity of demand (Fig. 10.6 illustrates the subscription price distribution). Thus, we calculated the average price per unit distance (a kilometre) for each month to eliminate the effect of travel distance based on information of the posted subscription price and the corresponding travel distance (shortest path distance), as it was commonly employed as an important factor influencing ridership in PT systems (Balcombe et al. 2004; Blainey and Preston 2010). The price per kilometre is the average service price level over a unit distance and therefore enables us to explore the price elasticity of CB demand.

For the station design of a CB route, in the ideal situation, all passengers board the CB at its origin station and alight at its terminal station. In practice, the origin and terminal stations are designed to coincide with the locations at which the largest numbers of passengers on a specific CB wish to board and alight. Passengers who

Table 10.1 Summary of variable descriptions

Variable	Description
Demand	Amount of CB demand
Characteristic of service supply	
Demand distance	Passenger travel distance in road network (km)
Price per km	Average subscription rate per kilometre (RMB/km)
Boarding station order	Order position of the boarding station for a passenger on his/her CB route
Alighting station order	Order position of the alighting station for a passenger on his/her CB route
Land use and accessibility	
Manufacturing	Category ratio of POIs belonging to industrial and manufacturing spaces
Administration	Category ratio of POIs belonging to administration and public services
Commerce	Category ratio of POIs belonging to commercial and business facilities
Green space	Category ratio of POIs belonging to green spaces and parks
Residence	Category ratio of residential POIs
Transportation	Category ratio of POIs belonging to roads, streets, and transportation spaces
Distance to CBD	Euclidean distance to the nearest CBD (km)
PB coverage	Total service area of public bus (PB) stations for a unit (km^2)
Railway coverage	Total service area of urban railway stations for a unit, including subway and rapid railway stations (km^2)
Net. connectivity	Index of connectivity to the road network for a unit
Land mix	Land-use mix index
Demographic characteristics	
Population 2010	Population within a spatial cell according to 2010 census, divided by 100,000 persons
GDP 2010	Total GDP within a spatial cell according to 2010 census, divided by 100 million RMB

board the CB at (or close to) the origin station and alight at (or close to) the terminal station are treated as the main demand flow, and their travel requests, such as their expected arrival times, are given top priority. The CB routes are more highly customized and suitable for these users and for the spatial units from which they are travelling. Thus, whether a specific cell is advantageous in the CB route design can be determined from its position in the ordering of stations by the boarding/alighting of passengers (see Fig. 10.7 for the illustration of station order). In essence, a lower average order position for a boarding station and a higher order position for an alighting station indicates an advantageous geographic position for a spatial unit (i.e., the unit is close to the origin and terminal stations of CB routes) in the CB route design, and it is consequently more likely to be associated with a higher CB demand.

Table 10.2 Descriptive statistics of variables (pooled by month and unit)

Variable (unit)	Mean	Std. Dev	Min	Max
Demand (person-time)	5.729	20.506	0	405
Demand distance (km)	17.133	10.023	2.968	46.979
Price per km (RMB/km)	0.645	0.048	0.6131	0.783
Boarding station order	4.048	2.384	1	13
Alighting station order	9.394	1.922	2	20
Manufacturing	0.024	0.042	0	0.180
Administration	0.224	0.178	0	1
Commerce	0.115	0.183	0	1
Green space	0.109	0.192	0	0.819
Residence	0.301	0.253	0	1
Transportation	0.226	0.192	0	0.953
Distance to CBD (km)	5.049	4.627	0.0945	24.404
PB coverage (km^2)	0.669	0.304	0	1
Railway coverage (km^2)	0.183	0.312	0	1
Net. connectivity	0.121	0.226	0	1.490
Land mix	0.612	0.249	0	0.946
Population 2010 (10^5 person)	0.054	0.058	0	0.235
GDP 2010 (10^8 RMB)	2.374	1.496	0.0002	7.993

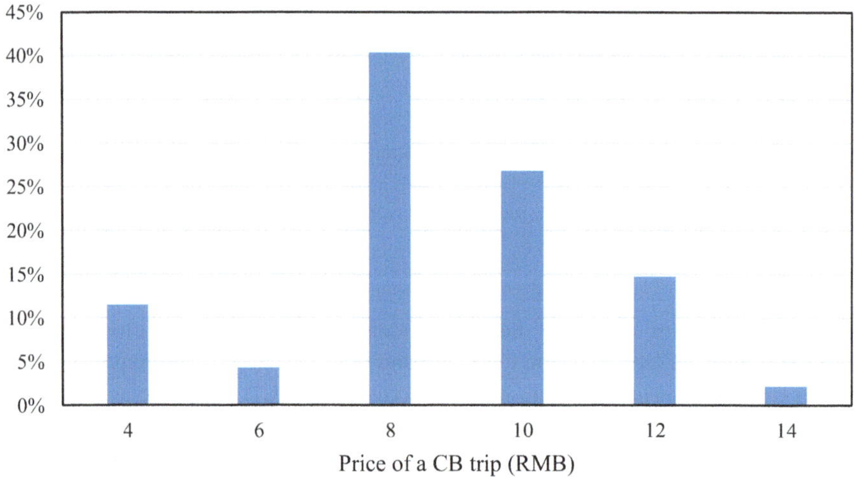

Fig. 10.6 Distribution of subscription prices of CB trips

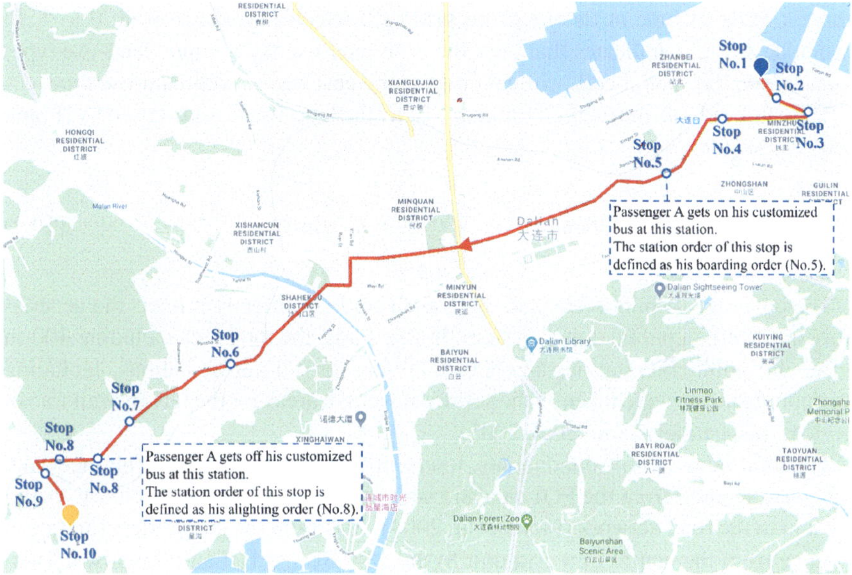

Fig. 10.7 An illustration of boarding/alighting station order (image based on Google Maps)

The attributes of land use and accessibility for the spatial units are not available from the observed subscription data set. It was therefore necessary to derive these attributes from other accessible geospatial big data sources. For example, information on points of interest (POIs) can be used to infer parcel functions because of their accessibility from online maps (Hu and Han 2019; Yao et al. 2017). The Dalian POI data set was obtained via the application programming interface of the Gaode Maps service in September 2018. In total, 325,217 POI records in 22 major categories were obtained and merged into six primary types based on the standard classification issued by ((MOHURD) Ministry of Housing and Urban–Rural Development of the People's Republic of China 2012). Specifically, this classification distinguishes POI types for residential areas, administration and public services, commercial and business facilities, industrial and manufacturing facilities, transportation services, and green spaces and parks. The POIs for administration and public services, commercial and business facilities, and industrial and manufacturing facilities identify places of employment, while the residential POIs indicate the homes of residents.

After screening, 250,201 valid POIs remained and were used to identify the land-use characteristics of the spatial units. The category ratio for each type of POI (Hu and Han 2019) within each grid unit was calculated to infer the land-use type:

$$C_m = \frac{\frac{n_m}{N_m}}{\sum_{l=1}^{6} \frac{n_l}{N_l}}, (m, l \in \{1, 2, \ldots, 6\}) \tag{10.1}$$

where n_m is the number of POIs for land-use type m within the grid unit; N_m is the total number of POIs for land-use type m in the entire city; and C_m is the category ratio for POIs of type m for the given grid cell, which may range from 0 to 1. In particular, $C_m = 1$ indicates that only the m-th land-use type (single land-use type) exists within the spatial cell, while 0 denotes the absence of such land-use type.

This study adopts the method for quantifying the land-use mix proposed by (Frank et al. 2004):

$$landmix = -\sum_{l=1}^{6} C_l \ln(C_l)/\ln(6) \tag{10.2}$$

The coverage of public bus and rail transit services directly reflects the accessibility of the urban public transit system. In this study, the total area including 400-m buffers for public bus stations and 800-m buffers for rail transit stations, excluding overlapping areas, were taken as the service coverage areas for the bus and rail transit services, respectively (Kim et al. 2005).

The total numbers of roundabouts, intersections, and expressway entrances and exits were fetched from the POI data and summed to reflect the connectivity (node) of a unit to the road network (Frank et al. 2004). Additionally, the shortest Euclidean distance from the centre of a grid unit to the 13 CBDs in Dalian (Qingniwa, Tianjinjie, Xianlu, Heping, Heishijiao, Wandagaoxin, Aolinpike, Dongte, Huananguanchang, Ganjingzilu, Quanshui, Bayilu, Kaifaquansheng) was calculated to represent accessibility of public city centres (Yang and Cao 2018).

As indicated in a previous study (Balcombe et al. 2004), GDP and population are good indicators of demographic characteristics. (Huang et al. 2014) and (Fu et al. 2014) generously shared their processed 1-km-scale population and GDP data sets, respectively, for China. The population and GDP for each cell are derived on the basis of the most recent population census of the People's Republic of China conducted in 2010 and correspond to the year 2010.

10.4.2 CB Demand Modelling

10.4.2.1 Spatial Dynamic Panel Data Model

In this study, we employed spatial dynamic panel data models to simultaneously accommodate dynamic adjustments, spatial dependence and spillover effects, and unit-specific heterogeneity based on longitudinal observations from the Dalian CB system.

Generally, endogenous and exogenous interaction effects together explain why observations from a particular spatial unit depend upon observations from other units (Elhorst 2014). Endogenous interaction effects reveal how the value of the dependent variable for a particular unit depends on the corresponding values for surrounding areas. Furthermore, if an explanatory variable in a spatial unit changes, not only will its own dependent variable change, but also the dependent variables in other units will change (Elhorst 2012) due to exogenous interaction effects.

We use the following general form of a dynamic model in space and time, which is similar to the specification proposed by (Elhorst 2001):

$$\begin{aligned} \mathrm{y}_{i,t} = \alpha + \tau \mathrm{y}_{i,t-1} + \delta \sum_{j=1}^{n} w_{ij}\mathrm{y}_{j,t} + \eta \sum_{j=1}^{n} w_{ij}\mathrm{y}_{j,t-1} + \sum_{k=1}^{K} x_{i,t,k}\beta_k \\ + \sum_{k=1}^{K}\sum_{j=1}^{n} w_{ij} z_{j,t,k}\gamma_k + \varepsilon_{it} \end{aligned} \tag{10.3}$$

$$\varepsilon_{it} = \mu_i + v_{it} i = 1, \ldots, n; t = 1, \ldots, T \tag{10.4}$$

$$E(\mu_i) = E(v_{it}) = E(\mu_i v_{it}) = 0 \tag{10.5}$$

where $\mathrm{y}_{i,t}$ is the CB demand from unit i in period t; $\mathrm{y}_{i,t-1}$ introduces temporal adjustments, and τ captures the series dependence of the dependent variable; $\sum_{j=1}^{n} w_{ij}\mathrm{y}_{j,t}$ is the spatial lag of the dependent variable at time t, and δ is the spatial dependence parameter; w_{ij} is the element in row i and column j of the spatial weight matrix $\boldsymbol{W}$; $\sum_{j=1}^{n} w_{ij}\mathrm{y}_{j,t-1}$ is the space–time lag of the dependent variable, and η is the corresponding parameter capturing the pattern of spatiotemporal diffusion; $x_{i,t,k}$. denotes the k-th independent variable, and the corresponding coefficient is denoted by β_k; $\sum_{j=1}^{n} w_{ij} z_{j,t,k}$ denotes the spatially lagged k-th independent variable (if the spatial lag of an independent variable is excluded from the model, z is replaced by 0), and γ_k is the corresponding coefficient; and the disturbance term ε_{it} has two orthogonal components, one representing unobserved individual heterogeneity (μ_i) and the other representing idiosyncratic shocks (v_{it}) (Roodman 2009a).

As discussed above, the inclusion of endogenous interactions, including $\sum_{j=1}^{n} w_{ij}\mathrm{y}_{j,t}$ and $\sum_{j=1}^{n} w_{ij}\mathrm{y}_{j,t-1}$, and the exogenous interaction $\sum_{j=1}^{n} w_{ij} z_{j,t,k}$ will help reveal the mechanisms of spatial dependence and spillover effects on CB demand. Previous studies have indicated that the incorporation of a spatially lagged dependent variable will introduce issues of endogeneity (Abreu et al. 2011). To obtain unbiased and consistent estimates, it is necessary to consider $\sum_{j=1}^{n} w_{ij}\mathrm{y}_{j,t}$ and $\sum_{j=1}^{n} w_{ij}\mathrm{y}_{j,t-1}$ as eogenous variables (Bouayad-Agha and Lionel 2010) and instrument them with their own lags. Fortunately, we have a relatively large T, 26 months in total, and therefore, a large number of instruments can be generated.

We employed the system generalized method of moments (GMM) estimator (Arellano and Bover 1995; Blundell and Bond 1998), which combines moment conditions for difference equations with instruments in levels and moment conditions for level equations with instruments in first differences, to estimate the CB demand models. We used the standard treatment of instrumenting for the endogenous variable (Rdman 2009a), which is based on lags of the instrumented variables. More specifically, for each period, all available lags of each endogenous variable were employed as instruments in the difference equation (for instance, $\{y_{i,t-2}, y_{i,t-3}, \ldots\}$. as instruments of $\Delta y_{i,t-1}$ in the difference equation, where $\Delta y_{i,t-1} = y_{i,t-1} - y_{i,t-2}$), and the contemporaneous first difference was employed as an instrument in the levels equation (for instance, $\{\Delta y_{i,t-1}\}$ as an instrument of $y_{i,t-1}$ in the levels equation).

In the standard, the uncollapsed form of the instrument matrix creates one instrument for each period, variable, and lag distance (Roodman 2009a). The number of instruments increases with the number of observations. However, an excessively large number of instruments (greater than the number of cross-sections, as is the case in this study) may result in the overfitting of endogenous variables and lead to biased inferences (Roodman 2009b). Fortunately, (Roodman 2009a) suggested limiting the number of instruments by collapsing the instrument matrix (combining the instruments in subsets) to avoid redundancy between different periods without missing information. For instance, we collapsed the instrument matrix from the form:

$$\begin{pmatrix} 0 & 0 & 0 & 0 & 0 & 0 & \ldots \\ y_{i1} & 0 & 0 & 0 & 0 & 0 & \ldots \\ 0 & y_{i1} & y_{i2} & 0 & 0 & 0 & \ldots \\ 0 & 0 & 0 & y_{i1} & y_{i2} & y_{i3} & \cdots \\ \vdots & \vdots & \vdots & \vdots & \vdots & \vdots & \ddots \end{pmatrix} \tag{10.6}$$

to the form:

$$\begin{pmatrix} 0 & 0 & 0 & \ldots \\ y_{i1} & 0 & 0 & \ldots \\ y_{i2} & y_{i1} & 0 & \ldots \\ y_{i3} & y_{i2} & y_{i1} & \cdots \\ \vdots & \vdots & \vdots & \ddots \end{pmatrix} \tag{10.7}$$

for the difference equation. Additionally, the instrument matrix for the levels equation:

$$\begin{pmatrix} 0 & 0 & 0 & \ldots \\ \Delta y_{i2} & 0 & 0 & \ldots \\ 0 & \Delta y_{i3} & 0 & \ldots \\ 0 & 0 & \Delta y_{i4} & \cdots \\ \vdots & \vdots & \vdots & \ddots \end{pmatrix} \tag{10.8}$$

was collapsed to a single column:

$$\begin{pmatrix} 0 \\ \Delta y_{i2} \\ \Delta y_{i3} \\ \Delta y_{i4} \\ \vdots \end{pmatrix} \quad (10.9)$$

Ensuring the exogeneity of instruments is a crucial assumption for the validity of instruments in the GMM estimator (Roodman 2009a). Two commonly used statistical tests are the Sargan test (Sargan 1958) and the Hansen J test (Hansen 1982). The former is not robust, but the effectiveness of the test does not weaken as the number of instruments increases. The latter is robust, but the effectiveness weakens as the number of instruments increases. After collapsing the instrument matrices, only 64–65 instruments were ultimately used in our models, much fewer than the number of panels (423). Obviously, the robust Hansen test is the natural choice for our models.

Cross-sectional dependence is introduced by the spatial weight matrix $\boldsymbol{W}$, which reveals the spatial relationships between units of analysis. Previous studies have indicated that the estimates derived from different spatial weight matrices are very similar (LeSage and Pace 2009). In this study, in accordance with the fundamental attributes of on-demand transit services, the CB demand from neighbours in each direction may have spatial effects on a given unit of analysis. Therefore, we adopt the isotropic queens-based spatial contiguity weight matrix (Lagona 2010) with each row normalized.

In addition to the regression model discussed above, possible alternative models are count data models (such as Poisson regression), censored data models (such as Tobit regression), and sample selection models (Heckman-type selection models). In practice, the demand in PT systems is commonly modelled as normally distributed; therefore, count data models are seldom employed (Graham et al. 2009; Tsai et al. 2014, 2012). Since cells within the service area that did not exhibit observed demand during the research period are included in our investigation, parameter estimates using a Tobit regression or a sample selection model (using a dichotomy index to indicate whether a cell is passed by the existing CB routes and then Y to represent the CB demand) should be more consistent (Tobin 1958). As this study aims to investigate the spatiotemporal dependence and spillover effects in CB demand, limited by the drawbacks of the models themselves (e.g., the Tobit model is very sensitive to the sample distribution and heteroscedasticity [Brown and Moffitt 1983]) and their developments in spatial modelling, it is extremely difficult to estimate the parameters for the general specification of the spatial dynamic panel model developed in this paper (i.e., to simultaneously accommodate dynamic adjustments, spatial dependence, spillover effects, and unit-specific heterogeneity).

10.4.2.2 Direct and Indirect Effects (Spatial Spillover Effects)

The CB demand model proposed above exploits the complicated structure of dependence among the spatial units of analysis. The estimated parameters reveal a wealth of information on the relationships between the observations and these spatial units. Unlike in the case of spatially independent data, the partial derivative of the dependent variable with respect to a certain explanatory variable may not be equal to its coefficient, which instead takes an adjusted value as indicated by the corresponding element of the matrix $(\boldsymbol{I}_n - \delta \boldsymbol{W})^{-1}$ (LeSage and Pace 2009). This implies that a change in the CB demand for a particular spatial unit, as associated with each explanatory variable, will affect the unit itself through a direct effect and may potentially influence other units indirectly. Such an indirect effect is also called a spatial spillover effect and is always a focus of empirical studies on applications of spatial econometric approaches (Elhorst 2012).

The CB demand model specified in Eq. (10.3) can be rewritten in vector form for a particular panel of observations at time point t as follows:

$$\boldsymbol{Y}_t = \alpha \boldsymbol{l}_n + \tau \boldsymbol{Y}_{t-1} + \delta \boldsymbol{W} \boldsymbol{Y}_t + \eta W Y_{t-1} + \boldsymbol{X}_t \beta + \boldsymbol{W}\boldsymbol{Z}_t \gamma + \boldsymbol{\varepsilon}_t \tag{10.10}$$

where $\boldsymbol{l}_n$ is an $n \times 1$ vector of ones, with α being the associated intercept parameter; $\boldsymbol{Y}_t$ is an $n \times 1$ vector consisting of one observation of CB demand for each grid unit $(i = 1, \ldots, n)$ at time t $(t = 1, \ldots, T)$; $\boldsymbol{X}_t$ is an $n \times K$ matrix of exogenous explanatory variables; $\boldsymbol{W}\boldsymbol{Z}_t$ is an $n \times K$ matrix of spatial lags of explanatory variables; and the $n \times 1$ vectors $\boldsymbol{\beta}$ and $\boldsymbol{\gamma}$ contain the parameters of the exogenous explanatory variables.

To assess short-term effects, τ and η in Eq. (10.6) are ignored. Thus, the matrix of the partial derivatives of $\boldsymbol{Y}$ with respect to the k-th explanatory variable of $\boldsymbol{X}$ for all spatial units at time t can be obtained as follows:

$$\left[\frac{\partial Y}{\partial x_{1k}} \cdots \frac{\partial Y}{\partial x_{nk}}\right]_t = (\boldsymbol{I}_n - \delta \boldsymbol{W})^{-1}(\beta_k \boldsymbol{I}_n + \gamma_k W) \tag{10.11}$$

where β_k is the k-th element of the vector $\boldsymbol{\beta}$, and γ_k is the k-th element of the vector $\boldsymbol{\gamma}$. . (LeSage and Pace 2009) proposed the following method for calculating the corresponding direct and indirect effects:

$$\text{short - term direct effect} = [(\boldsymbol{I}_n - \delta \boldsymbol{W})^{-1}(\beta_k \boldsymbol{I}_n + \gamma_k \boldsymbol{W})]^{\overline{d}} \tag{10.12}$$

$$\text{short - term indirect effect} = [(\boldsymbol{I}_n - \delta \boldsymbol{W})^{-1}(\beta_k \boldsymbol{I}_n + \gamma_k \boldsymbol{W})]^{\overline{rsum}} \tag{10.13}$$

where the superscript $\overline{d}$ denotes an operator that averages the row sums of the diagonal elements of a matrix, while the superscript $\overline{rsum}$ indicates averaging of the row sums of the off-diagonal elements of a matrix.

For the assessment of long-term effects, $\boldsymbol{Y}_t$ is assumed to be equal to $\boldsymbol{Y}_{t-1}$. In this case, the matrix of partial derivatives of $\boldsymbol{Y}$ with respect to the k-th explanatory variable is given by

$$\left[\frac{\partial Y}{\partial x_{1k}} \cdots \frac{\partial Y}{\partial x_{nk}}\right]_t = ((1-\tau)\boldsymbol{I}_n - (\delta+\eta)\boldsymbol{W})^{-1}(\beta_k \boldsymbol{I}_n + \gamma_k W) \quad (10.14)$$

Similarly, the direct and indirect effects over the long term are calculated as follows:

$$\text{long - term direct effect} = [((1-\tau)\boldsymbol{I}_n - (\delta+\eta)\boldsymbol{W})^{-1}(\beta_k \boldsymbol{I}_n + \gamma_k \boldsymbol{W})]^{\overline{d}} \quad (10.15)$$

$$\text{long - term indirect effect} = [((1-\tau)\boldsymbol{I}_n - (\delta+\eta)\boldsymbol{W})^{-1}(\beta_k \boldsymbol{I}_n + \gamma_k \boldsymbol{W})]^{\overline{rsum}} \quad (10.16)$$

10.5 Results and Discussion

The effects of factors related to the characteristics of CB service, demographic characteristics, land use and accessibility on CB demand were examined using spatial dynamic panel data models of the general form specified in Sect. 4.2. The full models with all the factors discussed in Sect. 4.1 were estimated first for an overview of their effects. The estimation results are summarized in the Appendix (Table 10.6).

In particular, $\boldsymbol{WY}_t$ indicates a contemporary spatial dependence effect, namely, an instant and rapid spatial diffusion effect, while the space–time lag $\boldsymbol{W}Y_{t-1}$ indicates a non-contemporary spatial dependence effect, namely, a slower diffusion towards the surrounding spaces) (Lopez and Chasco 2004). The coefficient estimates of contemporary and non-contemporary spatial diffusion (space–time diffusion) were tested to present a very similar magnitude of approximately 0.43 but opposite directions (as indicated by the full model estimation results listed in Table 10.6), which indicates $\delta = -\eta$ in Eq. (10.3). Therefore, the final model specification in vector form can be written as:

$$\boldsymbol{Y}_t = \alpha \boldsymbol{l}_n + \tau \boldsymbol{Y}_{t-1} + \delta \Delta \boldsymbol{WY}_t + \boldsymbol{X}_t \beta + \boldsymbol{WZ}_t \gamma + \boldsymbol{\varepsilon}_t \quad (10.17)$$

where $\Delta \boldsymbol{WY}_t = \boldsymbol{WY}_t - \boldsymbol{W}Y_{t-1}$, and other terms are the same as in Eq. (10.10). Additionally, variables with sufficiently small t-statistics were excluded (from Table 10.6) through backward elimination, and the estimation results of the final models are listed in Table 10.3.

As discussed in Sect. 4.1, subscription price per kilometer, averaged travel distance supplied, and boarding/alighting station order are employed as indicators of CB service supply and examined in Models 1, 2 and 3, respectively. The model choice was assessed based on the significance of the estimates, the Arellano-Bond test for

Table 10.3 Estimation results of CB demand models (robust t-statistics in parentheses)

	Model 1	Model 2	Model 3
	Coeff	Coeff	Coeff
$Demand_{t-1}$	0.983***	0.979***	0.968***
	(172.76)	(136.50)	(126.48)
$\Delta w.Demand_t$	0.442***	0.439***	0.436***
	(2.95)	(2.93)	(2.94)
Price per km	−2.736***		
	(−4.35)		
Demand distance		0.047***	
		(10.18)	
Order of boarding station			−0.318***
			(−2.83)
Order of alighting station			0.303***
			(4.51)
GDP 2010	0.097	0.106	0.118
	(1.18)	(1.25)	(1.28)
Distance to CBD	−0.020***	−0.020**	−0.011
	(−2.80)	(−2.49)	(−1.05)
Residence	0.352	0.376*	0.400
	(1.59)	(1.68)	(1.56)
Commerce	0.147	0.162	0.258*
	(1.29)	(1.40)	(1.81)
Net. Connectivity	0.139	−0.019	0.037
	(0.48)	(−0.06)	(0.10)
Land mix	0.929***	0.646***	0.419*
	(3.88)	(2.68)	(1.79)
PB coverage	0.007	−0.437	−0.769
	(0.01)	(−0.85)	(−1.24)
w.GDP 2010	−0.207**	−0.207**	−0.196*
	(−2.21)	(−2.18)	(−1.74)
w.Residence	0.475	0.436	0.535
	(0.91)	(0.83)	(0.88)
w.Commerce	0.600	0.367	0.758
	(0.99)	(0.61)	(1.06)
w.Net.Connectivity	−1.048***	−0.398	−0.281
	(−3.26)	(−1.25)	(−0.81)

(continued)

Table 10.3 (continued)

	Model 1	Model 2	Model 3
	Coeff	Coeff	Coeff
w.Landmix	0.447	0.195	0.716
	(0.84)	(0.35)	(0.98)
Intercept	1.269**	−0.187	−0.581
	(2.41)	(−0.60)	(−1.50)
No. of panels	423	423	423
No. of instruments	64	64	65
AR(2) test	−1.57	−1.57	−1.57
(AR(2) p-value)	0.12	0.12	0.12
Hansen test	53.31	51.10	49.96
(Hansen p-value)	0.28	0.35	0.40

Note (1) *w.x* denotes a spatial lag of *x*; ; (2) * $p < 0.1$, ** $p < 0.05$, *** $p < 0.01$

second-order serial correlation in disturbances (Arellano and Bond 1991) (denoted by AR(2) for short), and the Hansen test of over-identifying restrictions (Hansen 1982).

None of the three models can reject the null assumptions of the Hansen test (joint validity of all instruments) or the Arellano-Bond test (error terms are not serially correlated) at a 10% level of significance. These tests demonstrate that the specifications of our CB demand models are appropriate.

According to the parameter estimates, the CB demand has a stronger time dependence than spatial dependence. The coefficient estimat of $\Delta w.Demand_t$. are highly significant and in the range of 0.436–0.442, revealing the mechanism of spatial dependence such that the increments (or decrement) in demand rather than the amount of demand in the neighbourhood contribute to the spatial dependence effects in CB demand. In particular, a unit increase in the demand of neighbouring cells in the current month relative to the previous month ($\Delta w.Demand_t = w.Demand_t - w.Demand_{t-1}$) will result in a 0.436 (to 0.442)-unit increment in the demand of a given unit for this month ($Demand_t$). This outcome may be observed because either some new CB routes are launched or the existing routes are adjusted to serve new demand that then actively absorbs potential demand along the routes, especially within their origin/destination areas. In this case, there must be more than 43% (of the increment in the neighbourhood) of new demand in a given area matched with an increment of demand in the neighbourhood (weighted).

Under model specifications involving spatial dependence effects, the coefficients of the explanatory variables included in the model cannot be directly interpreted since the partial derivative with respect to each factor is not equal to the corresponding regression coefficient in the vector $\boldsymbol{\beta}$ (LeSage and Pace 2009). However, the indirect effects of the independent explanatory variables can be validly used to test the hypothesis of the existence of spatial spillover effects, rather than considering only the coefficient estimates of the exogenous interaction terms (Elhorst 2014).

Based on the coefficient estimates in Table 10.3, we assessed the direct and indirect effects of the explanatory variables via Eqs. (10.11)–(10.16). To infer the significance of these impacts, we drew at random from multivariate normal distributions with the means (the parameter estimates given in Table 10.3) and the variance–covariance matrices (estimated from the models) (LeSage and Pace 2009). The short-term direct and indirect effects are reported in Tables 10.4 and 10.5, respectively, with standard deviations derived from 1000 draws. The long-term direct and indirect effects are insufficiently significant and therefore not listed.

In terms of the characteristics of service supply, as expected, the subscription price per kilometre has a significantly negative direct effect on CB demand, which is consistent with findings for traditional PT systems (Graham et al. 2009; Paulley et al. 2006). A 0.1 RMB decrease in the price per km will increase the CB demand by 0.286 units, revealing a price elasticity of −0.32 at the means (which is tested robust

Table 10.4 Short-term direct effects (t-statistics in parentheses)

	Model 1	**Model 2**	**Model 3**
Price per km	-2.856^{***}		
	(−4.53)		
Demand distance		0.049^{***}	
		(9.36)	
Order of boarding station			-0.332^{**}
			(−2.43)
Order of alighting station order			0.316^{***}
			(3.65)
GDP 2010	0.081	0.09	0.104
	(1.12)	(1.12)	(1.18)
Distance to CBD	-0.021^{**}	-0.021^{**}	−0.012
	(−2.53)	(−2.4)	(−1.07)
Residence	0.415	0.436	0.469
	(1.43)	(1.49)	(1.30)
Commerce	0.213	0.205	0.344^{*}
	(1.36)	(1.38)	(1.68)
Net. Connectivity	0.041	−0.059	0.011
	(0.14)	(−0.25)	(0.03)
Land mix	1.015^{***}	0.693^{**}	0.506^{*}
	(3.08)	(2.33)	(1.72)
PB coverage	0.008	−0.456	−0.802
	(0.03)	(−0.81)	(−1.14)

Note * $p < 0.1$, $^{**} p < 0.05$, $^{***} p < 0.01$

Table 10.5 Short-term indirect (spillover) effects (t-statistics in parentheses)

	Model 1	Model 2	Model 3
Price per km	−2.048		
	(−1.19)		
Demand distance		0.035	
		(1.22)	
Boarding station order			−0.233
			(−0.76)
Alighting station order			0.222
			(0.86)
GDP 2010	−0.277	−0.271*	−0.242
	(−1.57)	(−1.79)	(−1.35)
Distance to CBD	−0.015	−0.015	−0.008
	(−0.71)	(−0.93)	(−0.62)
Residence	1.068	1.013	1.19
	(0.63)	(0.74)	(0.66)
Commerce	1.127	0.739	1.459
	(0.83)	(0.67)	(0.97)
Net. connectivity	−1.671*	−0.685	−0.444
	(−1.87)	(−1.17)	(−0.74)
Land mix	1.452	0.807	1.506
	(0.73)	(0.63)	(0.74)
PB coverage	0.005	−0.324	−0.562
	(0.20)	(−0.56)	(−0.57)

Note * $p < 0.1$, ** $p < 0.05$, *** $p < 0.01$

with a level-log transformation of Eq. [10.17]), which coincides with the elasticity of conventional public bus systems (Litman 2004; Paulley et al. 2006).

The significantly positive estimate of demand distance indicates that the CB service is more attractive for longer trips. This finding is consistent with previous studies that have also identified an increased focus on long-distance trips as a key characteristic of successful subscription bus services (Kirby and Bhatt 1975; Zhang et al. 2017). As intuited, this finding may be attributed to the fact that the long distance amplifies the passengers' perception of the improved service quality they experience when shifting from conventional PT services to CB services (Wang et al. 2020; Zhang et al. 2017) because most passengers are previous users of conventional PT services (Wang et al. 2019). Specifically, a one-kilometre increase in the average demand distance of an area will bring a 0.049-unit increase in demand, while the increase in the demand distance in other cells will also increase the demand in this cell (although not enough to be statistically significant). On the other hand, the longer

CB routes pass through more spatial units and therefore have more opportunity to absorb potential demand.

As discussed in Sect. 4.1, a unit with an earlier position in the boarding station order and a later order position in the alighting station order has an advantage in CB route design. The model estimates verify our intuition with a negative estimate of the boarding station order and a positive estimate of the alighting station order. According to the magnitude of direct effects, an earlier place in the boarding station order is slightly more effective in increasing the CB demand.

CB services, as an emerging luxury bus mode, are attracting more wealthy commuters (those living in areas with a higher GDP) compared to conventional PT services (Liu and Ceder 2015a; Zhang et al. 2017), as implied by the positive direct effects of GDP. GDP displays a negative spatial spillover effect on CB demand and is significant at the 90% statistical level in Model 2,[1] indicating that a one-unit increase in the GDP of its neighbours will induce an approximately 0.271-unit decrease in the demand of a given unit.

There is significantly more CB demand generated in units closer to a CBD. This outcome occurs because more people reside in or near CBDs, as indicated by a significant negative linear relationship between population and the distance to a CBD (with the coefficient estimated at −0.005 and the t-statistic at −42.19). The population level has a significant relationship with CB demand (Wang et al. 2019) but a non-significant (but still positive) relationship in the spatial dynamic CB demand models, possibly because it was assumed that the population was fixed during the research period. Consequently, a higher demand level is also associated with a greater proportion of residential and commercial areas, as shown by their direct effects. The proportions of residential and commercial areas have positive but non-significant spatial spillover effects on CB demand, respectively. Specifically, as white-collar workers dominate the passenger groups, residential and commercial areas (generally with more workplaces distributed) are the primary locations for CB stops.

The land-use pattern exerts a significantly positive direct effect on the CB demand, which is consistent with the findings from (Frank and Pivo 1994) that transit usage increases with the land-use mix. In addition, the CBD and population level significantly and positively affect the land-use mix level (Huang and Tsai 2013). The units close to the CBDs and with higher population levels also have higher land-use mix indexes and therefore are significantly associated with the higher CB demand level.

The connectivity of a unit exerts negative effects on CB demand in both direct and indirect ways, suggesting that there are more potential CB users from cells with poor connectivity to the road network. Poor connectivity to the road network is not conducive to the development of PT services and is not friendly to private car owners. Therefore, the CB services in these areas have lower substitutability (in terms of the mode of commuting).

[1] Even though the estimate for *GDP 2010* is not significant in Table 10.3, the coefficient for *w.GDP 2010* is highly significant. As illustrated in Eq. (10.13), they both contribute to the indirect effect (significant at the 90% level).

The negative effect of the coverage of public bus stations suggests that more CB demand is generated in areas with poorer public bus service, which is consistent with the conclusion of (Wang et al. 2019). These findings indicate that CB services have a large market potential in areas with poor accessibility (to PT services and road networks).

To examine whether there were significant differences in the demand evolution patterns at different stages ("rapid growth period" and "steady operation period"), we estimated the models with the post-December-2015 samples only (and all the other specifications kept). The results are summarized in Table 10.7. Then, a t-test (Clogg et al. 1995) has been performed to compare the coefficients of the two models (Table 10.3 and Table 10.7). The results (listed in Table 10.8) indicate that the coefficient estimates are not significantly different from the full sample model, except for the temporal dependence effects (which may be due to the number of instruments and their information decrease given some periods deleted).

10.6 Conclusions

This study provides empirical evidence of spatial dependence and spillover effects in CB demand based on continuous observations of the Dalian CB system over more than two years. The proposed CB demand models simultaneously accommodate dynamic adjustments, spatial dependence, spatial spillover effects, and unit-specific heterogeneity. The endogeneity caused by blending spatial dependence effects has been resolved by extending the moment restrictions of the system GMM estimator (Arellano and Bover 1995; Blundell and Bond 1998) to a spatial autoregressive dynamic panel. The key findings of this empirical analysis provide an effective basis for CB demand identification.

The results suggest a stronger time dependence than spatial dependence. The increments (or decrement) in demand rather than the amount of demand in the neighbourhood contribute to the significant spatial dependence effects in CB demand. In particular, an increment in the demand of neighbourhood in the current month relative to the previous month will generate approximately 43% of that increment in demand for a given unit in the current month. This finding can orient potential marketing based on the expectation that matching demand will be generated in an area when the demand (weighted value) in its neighbourhood increases.

As with other modes of transportation, a lower subscription price over a unit distance is negatively related to the CB demand. A 0.1 RMB decrease in the price will increase the CB demand by approximately 0.286 units. Consistent with a previous study (Zhang et al. 2017), CB services are found to be more attractive for passengers with long-distance trips. On the other hand, longer CB routes pass through more spatial areas and therefore have more opportunities to actively absorb potential demand. Thus, increasing the focus on long-distance trips is one possible strategy for increasing the ridership of CBs. In addition, spatial units where the passenger has a lower boarding order position and a higher alighting order position are significantly associated with a larger amount of CB demand.

Accessibility (represented by connectivity to the road network and PB service coverage) exerts negative effects on CB demand, indicating that CB services will have a larger market niche in areas with poor accessibility.

Appendix

See (Tables 10.6, 10.7, and 10.8)

Table 10.6 Estimation results for the full models (robust t-statistics in parentheses)

	Model 1	Model 2	Model 3
	Coeff	Coeff	Coeff
Demand$_{t-1}$	0.982***	0.978***	0.968***
	(178.68)	(137.48)	(132.2)
w.Demand$_t$	0.434***	0.434***	0.423***
	(2.92)	(2.92)	(2.89)
w.Demand$_{t-1}$	−0.440***	−0.436***	−0.434***
	(−2.99)	(−2.96)	(−2.96)
Price per km	−2.906***		
	(−3.92)		
Demand distance		0.049***	
		(10.30)	
Boarding station order			−0.329***
			(−2.82)
Alighting station order			0.312***
			(4.46)
Population 2010	0.535	0.333	−0.842
	(0.16)	(0.10)	(−0.23)
GDP 2010	0.087	0.097	0.097
	(0.93)	(1.00)	(0.94)
Railway coverage	0.207	0.009	−0.118
	(1.54)	(0.06)	(−0.73)
Distance to CBD	−0.047	0.037	0.088
	(−0.47)	(0.38)	(0.79)
Residence	0.463**	0.407*	0.355
	(2.04)	(1.74)	(1.37)
Administration	0.032	−0.011	−0.170
	(0.14)	(−0.04)	(−0.54)
Commerce	0.079	0.007	0.097
	(0.41)	(0.04)	(0.42)
Manufacturing	0.480	−0.356	−1.116
	(0.47)	(−0.32)	(−0.78)
Net. connectivity	0.063	−0.021	0.114
	(0.29)	(−0.09)	(0.35)
Land mix	0.928***	0.640***	0.439*
	(3.83)	(2.60)	(1.82)
PB coverage	−0.009	−0.403	−0.746
	(−0.02)	(−0.68)	(−1.06)
w.Population 2010	−1.314	0.901	2.498

(continued)

Table 10.6 (continued)

	Model 1	**Model 2**	**Model 3**
	Coeff	**Coeff**	**Coeff**
	(−0.33)	(0.22)	(0.65)
w.GDP 2010	−0.216*	−0.193*	−0.203
	(−1.93)	(−1.70)	(−1.56)
w.Railwaycoverage	−0.037	0.039	−0.076
	(−0.12)	(0.13)	(−0.23)
w.DistancetoCBD	0.018	−0.062	−0.109
	(0.19)	(−0.67)	(−1.00)
w.Residence	1.396*	1.456**	1.269
	(1.9)	(1.98)	(1.50)
w.Administration.	1.629**	1.833***	1.373*
	(2.28)	(2.64)	(1.80)
w.Commerce	1.254	1.055	1.347
	(1.51)	(1.26)	(1.39)
w.Manufacturing	4.709**	4.912**	1.891
	(2.28)	(2.)	(0.80)
w.Net.connectivity	−1.143***	−0.606*	−0.641
	(−3.20)	(−1.70)	(−1.43)
w.Landmix	0. 807	0.426	0.702
	(1.27)	(0.63)	(0.91)
w.PBcoverage	−0.003	−0.308	0.259
	(−0.01)	(−1.03)	(0.71)
Intercept	0.448	−1.072*	−1.169*
	(0.66)	(−1.92)	(−1.84)
No. of panels	423	423	423
No. of instruments	76	76	77
AR(2) test	−1.57	−1.57	−1.57
(AR(2) p-value)	0.117	0.116	0.116
Hansen test	58.31	56.36	56.36
(Hansen p-value)	0.170	0.219	0.219

Note (1) *w.x* denotes the spatial lag of *x*; ; (2) * $p < 0.1$, ** $p < 0.05$, *** $p < 0.01$

Table 10.7 Estimation results of CB demand models with post-December-2015 samples only (robust t-statistics in parentheses)

	Model 1	Model 2	Model 3
	Coeff	Coeff	Coeff
$Demand_{t-1}$	0.790***	0.806***	0.732***
	(13.44)	(13.20)	(14.01)
$\Delta w.Demand_t$	0.189**	0.196***	0.195**
	(2.55)	(2.62)	(2.55)
Price per km	−6.481*		
	(−1.80)		
Demand distance		0.069***	
		(2.69)	
Order of boarding station			−0.734*
			(−1.76)
Order of alighting station			0.704**
			(2.24)
GDP 2010	0.352	0.332	0.459
	(0.79)	(0.78)	(0.92)
Distance to CBD	−0.124*	−0.112	−0.118*
	(−1.77)	(−1.64)	(−1.73)
Residence	1.572	1.468	1.908
	(1.12)	(1.10)	(1.25)
Commerce	0.867*	0.821*	1.289**
	(1.70)	(1.77)	(2.05)
Net. Connectivity	0.438	0.152	0.263
	(0.34)	(0.14)	(0.18)
Land mix	4.067*	3.191	3.265**
	(1.90)	(1.64)	(2.15)
PB coverage	0.051	−0.683	−1.951
	(0.03)	(−0.34)	(−0.65)
w.GDP 2010	−0.943	−0.848	−1.043
	(−1.34)	(−1.26)	(−1.43)
w.Residence	2.687	2.36	3.201
	(0.88)	(0.84)	(0.96)
w.Commerce	3.552	2.821	4.421
	(1.15)	(1.02)	(1.23)
w.Net.Connectivity	−5.168*	−3.578	−3.825
	(−1.70)	(−1.39)	(−1.62)
w.Landmix	3.019	2.321	4.087
	(0.96)	(0.82)	(1.06)
Intercept	1.901	−1.481	−2.71

(continued)

Table 10.7 (continued)

	Model 1	**Model 2**	**Model 3**
	Coeff	**Coeff**	**Coeff**
	(0.94)	(−0.96)	(−1.33)
No. of panels	423	423	423
No. of instruments	56	56	57
AR(2) test	−0.64	−0.65	−0.77
(AR(2) *p*-value)	0.522	0.516	0.444
Hansen test	50.01	68.15	58.14
(Hansen *p*-value)	0.09	0.004	0.032

Note (1) *w.x* denotes a spatial lag of *x*; ; (2) * $p < 0.1$, ** $p < 0.05$, *** $p < 0.01$

Table 10.8 A summary of t-values of differences in coefficient estimates between the models with post-December-2015 samples and full samples

	Model 1	Model 2	Model 3
$Demand_{t-1}$	−3.25	−2.82	−4.49
$\Delta w.Demand_t$	−1.51	−1.45	−1.45
Price per km	−1.02		
Demand distance		0.83	
Order of boarding station			−0.96
Order of alighting station			1.25
GDP 2010	0.56	0.52	0.67
Distance to CBD	−1.48	−1.34	−1.55
Residence	0.86	0.81	0.97
Commerce	1.38	1.38	1.60
Net. Connectivity	0.23	0.15	
Land mix	1.45	1.30	1.85
PB coverage	0.02	−0.12	−0.39
w.GDP 2010	−1.04	−0.94	−1.15
w.Residence	0.72	0.67	0.79
w.Commerce	0.94	0.87	1.00
w.Net.Connectivity	−1.35	−1.23	−1.49
w.Landmix.	0.81	0.74	0.86
Intercept	0.30	−0.82	−1.02

References

Abreu, M., H.L.F. de Groot, and R.J.G.M. Florax. 2011. Space and growth: A survey of empirical evidence and methods. *SSRN Electronic Journal* 21: 13–44. https://doi.org/10.2139/ssrn.631007.

Arellano, M., and S. Bond. 1991. Some tests of specification for panel data: Monte Carlo evidence and an application to employment equations. *Review of Economic Studies* 58: 277–297.

Arellano, M., and O. Bover. 1995. Another look at the instrumental variable estimation of error-components models. *J Econom* 68: 29–51.

Balcombe, R., R. Mackett, N. Paulley, J. Preston, J. Shires, H. Titheridge, M. Wardman, and P. White. 2004. *The demand for public transport: A practical guide, TRL limited*. Workingham: United Kingdom.

Becker, H., F. Ciari, and K.W. Axhausen. 2017. Modeling free-floating car-sharing use in Switzerland: A spatial regression and conditional logit approach. *Transp Res Part C Emerg Technol* 81: 286–299. https://doi.org/10.1016/j.trc.2017.06.008.

Blainey, S.P., and J.M. Preston. 2010. Modelling local rail demand in South Wales. *Transportation Planning and Technology* 33: 55–73. https://doi.org/10.1080/03081060903429363.

Blundell, R., and S. Bond. 1998. Initial conditions and moment restrictions in dynamic panel data models. *J Econom* 87: 115–143.

Boisjoly, G., E. Grisé, M. Maguire, M.P. Veillette, R. Deboosere, E. Berrebi, and A. El-Geneidy. 2018. Invest in the ride: A 14 year longitudinal analysis of the determinants of public transport ridership in 25 North American cities. *Transp Res Part A Policy Pract* 116: 434–445. https://doi.org/10.1016/j.tra.2018.07.005.

Bouayad-Agha, S., and V. Lionel. 2010. Estimation strategies for a spatial dynamic panel using gmm. A new approach to the convergence issue of European regions. *Spatial Economic Analysis* 5: 205–227. https://doi.org/10.1080/17421771003730711.

Bresson, G., J. Dargay, J.L. Madre, and A. Pirotte. 2003. The main determinants of the demand for public transport: A comparative analysis of England and France using shrinkage estimators. *Transp Res Part A Policy Pract* 37: 605–627. https://doi.org/10.1016/S0965-8564(03)00009-0.

Brown, C.C., and R.A. Moffitt. 1983. The effect of ignoring heteroscedasticity on estimates of the Tobit model. NBER technical paper series. National Bureau of Economic Research Cambridge, Mass., USA.

Chakrabarti, S. 2017. How can public transit get people out of their cars? An analysis of transit mode choice for commute trips in Los Angeles. *Transp Policy (Oxf)* 54: 80–89. https://doi.org/10.1016/j.tranpol.2016.11.005.

Chen, C., and C.E. McKnight. 2007. Does the built environment make a difference? Additional evidence from the daily activity and travel behavior of homemakers living in New York City and suburbs. *Journal of Transport Geography* 15: 380–395. https://doi.org/10.1016/j.jtrangeo.2006.11.010.

Clogg, C.C., E. Petkowa, and A. Haritou. 1995. Statistical methods for comparing regression coefficients between models. *American Journal of Sociology* 100: 1261–1293.

Dalian Municipal Bureau of Statistics. 2017. Dalian Statistic Yearbook 2017 [WWW Document]. URL http://www.stats.dl.gov.cn/index.php?m=content&c=index&a=lists&catid=57. Accessed 31 March 2018.

Dargay, J.M., and M. Hanly. 2002. The demand for local bus services in England. *Journal of Transport Economics and Policy* 36: 73–91.

de Rus, G. 1990. Public transport demand elasticities in Spain. *Journal of Transport Economics and Policy* 24: 189–201.

Elhorst, J.P. 2001. Dynamic models in space and time. *Geographical Analysis* 33: 119–140.

Elhorst, J.P. 2012. Dynamic spatial panels: Models, methods, and inferences. *Journal of Geographical Systems* 14: 5–28. https://doi.org/10.1007/s10109-011-0158-4.

Elhorst, J.P. 2014. *Spatial econometrics: From cross-sectional data to spatial panels*. Heidelberg: Springer.

Elhorst, J.P., 2017. *Spatial panel data analysis. In Encyclopedia of GIS*, 2050–2058. Cham: Springer International Publishing. https://doi.org/10.1007/978-3-319-23519-6_1641-1

FitzRoy, F., and I. Smith. 1998. Public transport demand in Freiburg: Why did patronage double in a decade? *Transp Policy (Oxf)* 5: 163–173. https://doi.org/10.1016/S0967-070X(98)00024-9.

Frank, L.D., M.A. Andresen, and T.L. Schmid. 2004. Obesity relationships with community design, physical activity, and time spent in cars. *American Journal of Preventive Medicine* 27: 87–96. https://doi.org/10.1016/j.amepre.2004.04.011.

Frank, L.D., and G. Pivo. 1994. Impacts of mixed use and density on utilization of three modes of travel: Single-occupant vehicle, transit, and walking. *Transportation Research Record: Journal of the Transportation Research Board* 1446: 44–52.

Friedrich, M., M. Hartl, and C. Magg. 2018. A modeling approach for matching ridesharing trips within macroscopic travel demand models. *Transportation (Amst)* 45: 1639–1653. https://doi.org/10.1007/s11116-018-9957-5.

Friedrich, M., and K. Noekel. 2017. Modeling intermodal networks with public transport and vehicle sharing systems. *EURO Journal on Transportation and Logistics* 6: 271–288. https://doi.org/10.1007/s13676-015-0091-7.

Fu, J., D. Jiang, Y. Huang. 2014. *Gridded population dataset of China (1km). Change research data publishing and repository*. https://doi.org/10.3974/geodb.2014.01.06.v1

García-Ferrer, A., M. Bujosa, A. de Juan, and P. Poncela. 2006. Demand forecast and elasticities estimation of public transport. *Journal of Transport Economics and Policy* 40: 45–67.

Graham, D.J., A. Crotte, and R.J. Anderson. 2009. A dynamic panel analysis of urban metro demand. *Transp Res E Logist Transp Rev* 45: 787–794. https://doi.org/10.1016/j.tre.2009.01.001.

Hansen, L.P., 1982. Large sample properties of generalized method of moments estimators. *Econometrica*: 1029–1054.

Hsiao, C. 2005. Why panel data? *Singapore Economic Review* 50: 143–154. https://doi.org/10.1142/S0217590805001937.

Hu, Y., and Y. Han. 2019. Identification of urban functional areas based on POI Data: A case study of the Guangzhou economic and technological development zone. *Sustainability (Switzerland)* 11: 1385. https://doi.org/10.3390/su11051385.

Huang, S.-W., and W.-J. Tsai. 2013. The analysis of measurements and influence factors of mixed land use. *Int J Biosci Biochem Bioinforma* 3: 206–210. https://doi.org/10.7763/ijbbb.2013.v3.197.

Huang, Y., D. Jiang, and J. Fu. 2014. 1 km grid GDP dataset of China (2005, 2010). *Global Change Research Data Publishing and Repository*. https://doi.org/10.3974/geodb.2014.01.07.v1.

Kain, J.F. and Z. Liu. 1999. Secrets of success: Assessing the large increases in transit ridership achieved by Houston and San Diego transit providers. *Transportation Research Part C: Emerging Technologies* 33: 601–624. https://doi.org/10.1016/S0965-8564(99)00009-9

Ke, J., H. Zheng, H. Yang, X. Chen. (Michael). 2017. Short-term forecasting of passenger demand under on-demand ride services: A spatio-temporal deep learning approach. *Transportation Research Part C: Emerging Technologies* 85: 591–608. https://doi.org/10.1016/j.trc.2017.10.016

Kim, Jumsan, Jongmin Kim, M. Jun, and S. Kho. 2005. Determination of a bus service coverage area reflecting passenger attributes. *Journal of the Eastern Asia Society for Transportation Studies* 6: 529–543.

Lagona, F. 2010. Model selection in Markov random fields for high spatial resolution hyperspectral data. In *Handbook of Applied Spatial Analysis*, 545–563. Springer.

LeSage, J., and R.K. Pace. 2009. *Introduction to spatial econometrics*. Chapman and Hall/CRC.

Litman, T. 2004. Transit price elasticities and cross-elasticities. *J Public Trans* 7: 37–58. https://doi.org/10.5038/2375-0901.7.2.3.

Liu, T., and A. Ceder. 2015. Analysis of a new public-transport-service concept: Customized bus in China. *Transp Policy (Oxf)* 39: 63–76. https://doi.org/10.1016/j.tranpol.2015.02.004.

Liu, T., A.A. Cedera, R. Bologna, and B. Cabantous. 2016. Commuting by customized bus: A comparative analysis with private car and conventional public transport in two cities. *J Public Trans* 19: 55–74. https://doi.org/10.5038/2375-0901.19.2.4.

Liu, T.L.K., P. Krishnakumari, and O. Cats. 2019. Exploring demand patterns of a ride-sourcing service using spatial and temporal clustering. 6th International Conference on Models and Technologies for Intelligent Transportation Systems. https://doi.org/10.1109/MTITS.2019.8883312

Lopez, F., and C. Chasco. 2004. *Space-time lags: Specification strategy in spatial regression models.* Champaign: Citeseer.

(MOHURD) Ministry of Housing and Urban-Rural Development of the People's Republic of China. 2012. *Code for classification of urban land use and planning standards of development land.*

National Bureau of Statistics of the People's Republic of China. 2011. Tabulation on the 2010 Population Census of the People's Republic of China [WWW Document]. URL http://www.stats.gov.cn/tjsj/pcsj/rkpc/6rp/indexch.htm. Accessed 31 March 2018

Linglin Ni, X. (Cara) Wang, X. (Michael) Chen. 2018. A spatial econometric model for travel flow analysis and real-world applications with massive mobile phone data. *Transportation Research Part C: Emerging Technologies* 86: 510–526. https://doi.org/10.1016/j.trc.2017.12.002

Pasha, M., S.M. Rifaat, R. Tay, and A. De Barros. 2016. Effects of street pattern, traffic, road infrastructure, socioeconomic and demographic characteristics on public transit ridership. *KSCE Journal of Civil Engineering* 20: 1017–1022. https://doi.org/10.1007/s12205-016-0693-6.

Paulley, N., R. Balcombe, R. Mackett, H. Titheridge, J. Preston, M. Wardman, J. Shires, and P. White. 2006. The demand for public transport: The effects of fares, quality of service, income and car ownership. *Transp Policy (Oxf)* 13: 295–306. https://doi.org/10.1016/j.tranpol.2005.12.004.

Roodman, D. 2009a. How to do xtabond2: An introduction to difference and system GMM in Stata. *Symposium (International) on Combustion* 9: 86–136.

Roodman, D. 2009b. Practitioners' corner: A note on the theme of too many instruments. *Oxford Bulletin of Economics and Statistics* 71: 135–158. https://doi.org/10.1111/j.1468-0084.2008.00542.x.

Sargan, J.D., 1958. The estimation of economic relationships using instrumental variables. *Econometrica*, 393–415.

Taylor, B.D., C.N.Y. Fink, E. Org. 2003. *Influencing transit ridership: A review and analysis of the ridership literature.*

Tobin, J. 1958. Estimation of relationships for limited dependent variables. *Econometrica*, 24–36.

Lu (Carol) Tong, L. Zhou, J. Liu, X. Zhou. 2017. Customized bus service design for jointly optimizing passenger-to-vehicle assignment and vehicle routing. *Transportation Research Part C: Emerging Technologies* 85, 451–475. https://doi.org/10.1016/j.trc.2017.09.022

Tsai, C.H., C. Mulley, and G. Clifton. 2012. The spatial interactions between public transport demand and land use characteristics in the Sydney Greater Metropolitan Area. *Road and Transport Research* 21: 62–73.

Tsai, C.H., C. Mulley, and G. Clifton. 2014. Forecasting public transport demand for the Sydney Greater Metropolitan Area: A comparison of univariate and multivariate methods. *A Journal of Australian and New Zealand Research and Practice* 23: 51–68.

Wang, J., T. Yamamoto, and K. Liu. 2019. Role of customized bus services in the transportation system: Insight from actual performance. *Journal of Advanced Transportation* 2019: 6171532. https://doi.org/10.1155/2019/6171532.

Wang, J., T. Yamamoto, and K. Liu. 2020. Key determinants and heterogeneous frailties in passenger loyalty toward customized buses: An empirical investigation of the subscription termination hazard of users. *Transp Res Part C Emerg Technol* 115: 102636. https://doi.org/10.1016/j.trc.2020.102636.

Yang, S., Y. Fan, W. Deng, and L. Cheng. 2019. Do built environment effects on travel behavior differ between household members? A case study of Nanjing, China. *Transp Policy (Oxf)* 81: 360–370. https://doi.org/10.1016/j.tranpol.2017.12.006.

Yang, W., and X. Cao. 2018. Examining the effects of the neighborhood built environment on CO2 emissions from different residential trip purposes: A case study in Guangzhou, China. *Cities* 81: 24–34. https://doi.org/10.1016/j.cities.2018.03.009.
Yao, Y., X. Li, X. Liu, P. Liu, Z. Liang, J. Zhang, and K. Mai. 2017. Sensing spatial distribution of urban land use by integrating points-of-interest and Google Word2Vec model. *International Journal of Geographical Information Science* 31, 825–848. https://doi.org/10.1080/13658816.2016.1244608
Yao, H., F. Wu, J. Ke, X. Tang, Y. Jia, S. Lu, P. Gong, J. Ye, and Z. Li. 2018. Deep multi-view spatial-temporal network for taxi demand prediction. 32nd AAAI Conference on Artificial Intelligence, AAAI 2018.
Zhang, J., D.Z.W. Wang, and M. Meng. 2017. Analyzing customized bus service on a multimodal travel corridor: An analytical modeling approach. *Journal of Transportation Engineering, Part A: Systems*. https://doi.org/10.1061/jtepbs.0000087

Uncited references

American Association of State Highway and Transportation Officials (AASHTO). Commuting in America 2021: The national report on commuting patterns and trends. Available from: https://traveltrends.transportation.org/wp-content/uploads/sites/62/2021/04/CA01-5.pdf
Chen, E., Z. Ye, and H. Wu. 2021. Nonlinear effects of built environment on intermodal transit trips considering spatial heterogeneity. *Transp Res D Transp Environ* 90: 102677. https://doi.org/10.1016/j.trd.2020.102677.
Eluru, N., and S. Yasmin. 2015. A note on generalized ordered outcome models. *Anal Methods Accid Res* 8: 1–6. https://doi.org/10.1016/j.amar.2015.04.002.
Liu, J., B. Wang, and L. Xiao. 2021. Non-linear associations between built environment and active travel for working and shopping: An extreme gradient boosting approach. *Journal of Transport Geography* 92: 103034. https://doi.org/10.1016/j.jtrangeo.2021.103034.
Liu, T., and A. Ceder. 2015b. Analysis of a new public-transport-service concept: Customized bus in China. *Transp Policy (Oxf)* 39: 63–76. https://doi.org/10.1016/J.TRANPOL.2015.02.004.
Ma, J., Y. Yang, W. Guan, F. Wang, T. Liu, W. Tu, and C. Song. 2017. Large-scale demand driven design of a customized bus network: A methodological framework and beijing case study. *Journal of Advanced Transportation* 2017. https://doi.org/10.1155/2017/3865701.
Meni, K., Avraham. N, K., and Mordechai, R., 1995. Commuting Stress: Causes, Effects, and Methods of Coping. Spring Science & Business Media.
Schlüter, J., A. Bossert, P. Rössy, and M. Kersting. 2021. Impact assessment of autonomous demand responsive transport as a link between urban and rural areas. *Research in Transportation Business & Management* 39: 100613. https://doi.org/10.1016/J.RTBM.2020.100613.

The manufacturer's authorised representative in the EU is Springer Nature Customer Service Centre GmbH, Europaplatz 3, 69115 Heidelberg, Germany. If you have any concerns regarding our products, please contact ProductSafety@springernature.com

Printed and bound by CPI Group (UK) Ltd, Croydon, CR0 4YY
01/10/2026
02225936-0004